The Politics of Low-Carbon Innovation

Per Ove Eikeland • Jon Birger Skjærseth

The Politics of Low-Carbon Innovation

The EU Strategic Energy Technology Plan

Per Ove Eikeland
Fridtjof Nansen Institute
Lysaker, Norway

Jon Birger Skjærseth
Fridtjof Nansen Institute
Lysaker, Norway

ISBN 978-3-030-17912-0   ISBN 978-3-030-17913-7   (eBook)
https://doi.org/10.1007/978-3-030-17913-7

Cover illustration: Melisa Hasan

This Palgrave Macmillan imprint is published by the registered company Springer Nature Switzerland AG
The registered company address is: Gewerbestrasse 11, 6330 Cham, Switzerland

PREFACE AND ACKNOWLEDGEMENTS

The EU Strategic Energy Technology Plan (SET-Plan), adopted in 2008, was aimed at stimulating and accelerating innovation in low-carbon energy technologies. The Commission presented the SET-Plan as the EU's main low-carbon energy technology-push policy instrument for achieving short- and long-term targets on decarbonization. It envisaged the Europeanization of a policy field where EU institutions, member-states, the research community and industry have varying strategic research and innovation interests—so political conflict was only to be expected.

Scant scholarly attention has been paid to explaining how the Plan came about, how it was filled with content and how it has worked in practice. At the ten-year anniversary of the SET-Plan, this book contributes to filling the void by adding analysis of technology policy to broader EU climate and energy policies.

Several scholars have provided valuable inputs and constructive criticism. Special thanks to our colleague at the Fridtjof Nansen Institute, Torbjørg Jevnaker; to Arne Fevolden at the Nordic Institute for Studies in Research, Innovation and Education (NIFU); and to Olav Wicken at the University of Oslo's TIK Centre for Technology, Innovation and Culture. Many scholars have also provided valuable feedback from presentations to international audiences along the way. We are as always grateful to Susan Høivik, who has improved the English text considerably. Further, we wish to thank representatives from EU institutions and national governments as well as interest organizations, for their openness and for taking the time to talk with us. Finally, we acknowledge our gratitude to the Research

Council of Norway's ENERGIX programme and the NordForsk project New Nordic Ways to Green Growth (NOWAGG) for granting funds for the project underlying this book.

Lysaker, Norway   Per Ove Eikeland
December 2018   Jon Birger Skjærseth

CONTENTS

ABOUT THE AUTHORS

Per Ove Eikeland is a senior research fellow at Fridtjof Nansen Institute. His research interests include European energy and climate policies, national energy transitions and corporate strategies. Main publications include co-edited and co-authored books: *Corporate Responses to EU Emissions Trading* (2013) and *Linking EU Climate and Energy Policies* (2016).

Jon Birger Skjærseth is research professor at Fridtjof Nansen Institute. His research interests include international environmental cooperation, EU climate and energy policies and corporate strategies. His articles and books have appeared in several publications. Among them are the following books which he co-edited and co-authored: *Corporate Responses to EU Emissions Trading* (2013) and *Linking EU Climate and Energy Policies* (2016).

ACRONYMS AND ABBREVIATIONS

AEBIOM	European Biomass Association
AGE	Advisory Group on Energy
ALTENER	Programme for the promotion of renewable energy sources in the European Union
BusinessEurope	Confederation of European business
CAN	Climate Action Network
CCS	Carbon capture and storage
CCS/U	Carbon capture, storage and use
CCU	Carbon capture and use
CEECs	Central and East European member-states of the European Union
CEPS	Centre for European Policy Studies
CIP	Competitiveness and Innovation Framework Programme of the European Union
Climate Alliance	Climate Alliance of European Cities with Indigenous Rainforest Peoples
COGEN Europe	European Association for the Promotion of Cogeneration
COM	Communication (from the European Commission)
Commission	European Commission
Council	Council of the European Union
CSP	Concentrated Solar Power
CSTP	Council for Science and Technology Policy (Japan)
DG	Directorate-General of the European Commission
EC	European Community
ECCP	European Climate Change Programme
ECSC	European Coal and Steel Community
EEC	European Economic Community

EEPR	European Energy Programme for Recovery
EERA	European Energy Research Alliance
EFSI	European Fund for Strategic Investments
EGEC	European Geothermal Energy Council
EIB	European Investment Bank
EII	European Industrial Initiative
ENTSO-E	European Network of Transmission System Operators for Electricity
ENVI	Environment, Public Health and Food Safety Committee of the European Parliament
EP	European Parliament
EPIA	European Photovoltaic Industry Association (from 2015: SolarPower Europe)
EPPSA	European Power Plant Suppliers Association
ERA	European Research Area
ERA-NET	Programme to spur Cooperation and Coordination between National Research Programmes in the European Union
ERAWOG	European Energy Research Area Working Group
ESR	Effort Sharing Regulation
ESTI	European Solar Test Installation
ETIP	European Technology and Innovation Platform
ETP	European Technology Platform
ETS	Emissions Trading System
EUA	European University Association
EURACOAL	European Association for Coal and Lignite
EURATOM	European Atomic Energy Community
EURELECTRIC	Union of the Electricity Sector
EuroHORCS	European Heads of Research Councils
EWEA	European Wind Energy Association
FORATOM	European Atomic Forum—Association for European Nuclear Industry
FP	Framework Programme for Research and Technological Development
GE	General Electric
Gen IV	Generation IV Fission Reactors
GHG	Greenhouse Gas
HORIZON 2020	8th Framework Programme for Research and Innovation of the European Union
ICF	International Company Engaged in Energy Consultancy
IEA	International Energy Agency

IEE	Intelligent Energy for Europe Programme of the European Union
IPCC	Intergovernmental Panel on Climate Change
ITRE Committee	Industry, Trade, Research and Energy Committee of the European Parliament
JOULE	Programme for Research and Development in the Field of Non-Nuclear Energy in the European Union
JRC	Joint Research Centre of the European Commission
JTI	Joint Technology Initiative
KIC	Knowledge and Innovation Community
LI	Liberal Intergovernmentalism
LULUCF	Land use, and Land-use Change and Forestry
MEP	Member of the European Parliament
MLG	Multilevel Governance
MS	Member-State
NER 300	EU Programme for Funding Low-Carbon Demonstration Projects
NER	New Entrants Reserve
OECD	Organisation for Economic Cooperation and Development
OJ	Official Journal of the European Union
OMC	Open Method of Coordination
ORGALIME	European Engineering Industries Association
PV	Photovoltaic
R&D	Research & Development
R&I	Research & Innovation
REN21	The Renewable Energy Policy Network for the 21st Century
RSFF	Risk-Sharing Finance Facility
SAVE	EU Programme for Specific Actions for Vigorous Energy Efficiency
SEC	Single European Code
SETIS	Strategic Energy Technology Information System
SET-Plan	Strategic Energy Technology Plan
SMEs	Small- and Medium-Sized Enterprises
SNETP	Sustainable Nuclear Energy Technology Platform
THERMIE	Programme for the Promotion of New and Innovative Energy Technologies in the European Union (demonstration activities)

LIST OF TABLES

CHAPTER 1

Introduction

This book examines the EU's low-carbon research and innovation (R&I) policy, launched with the Strategic Energy Technology Plan (SET-Plan) in 2008.[1] The Plan aimed at prioritizing and raising R&I funding, to cut costs for some promising energy technologies and project types while down-emphasizing others.[2]

Such a focus on certain technological priorities could redistribute future economic opportunities for various sectors, companies, research institutions and member-states. The Plan signalled stronger EU integration, challenging established national prerogatives to decide energy R&I. It also impinged on the principle of technology neutrality, tasking the market to pick winners. All this made the SET-Plan contested, in turn spurring mobilization and political conflict.

The SET-Plan was adopted as part of a larger package of climate and energy policies. This package kick-started the EU's fight against global climate change and growing energy-import dependency, while it also aimed at boosting the international competitiveness of European

---

[1] 'Technological innovations' comprise new products and processes and significant technological changes of products and processes. An innovation has been 'implemented' if it has been introduced in the market (OECD 2013).

[2] The SET-Plan envisaged specific priority to supporting and realizing large demonstration projects, viewed in the innovation literature as critical for realizing ideas as innovations for the market (Bossink 2015; IEA 2015). Demonstration projects typically test full-scale workability of the technology to prove its utility to potential users in the market (e.g. see Myers 1978).

© The Author(s) 2020   1
P. O. Eikeland, J. B. Skjærseth, *The Politics of Low-Carbon Innovation*, https://doi.org/10.1007/978-3-030-17913-7_1

businesses. Together with interim targets for 2020,[3] a long-term target was set: to reduce EU greenhouse gas (GHG) emissions by 80–95 per cent by 2050. Several EU directives and regulations were adopted, including the Renewable Energy Directive and the revised EU Emissions Trading Directive, to *pull* market deployment of new low-carbon solutions. The SET-Plan became the complementary technology *push* pillar, aimed at accelerating innovation through the development and demonstration of technology that would lower costs. Together, push and pull policies would shape the EU's collective capacity for low-carbon technology innovation and deployment. The package was unprecedented—never had such ambitious EU policies been adopted to stimulate demand and supply of low-carbon technologies.

While EU climate and energy policies have received considerable scholarly attention,[4] EU low-carbon technology policies and the politics of innovation have remained largely unexplored terrain.[5] This seems puzzling, given the saliency of the SET-Plan for achieving EU decarbonization, as underscored by the Commission: 'Energy technology holds the key to success. Without a sustained and increased effort, the targets will not be met' (Commission 2007, p. 4).

This book takes a first step towards filling this knowledge gap. At ten-year anniversary of the SET-Plan, we take stock of its development, from initiation to the current state of implementation, asking: How was the SET-Plan established? Have implementation and performance been in line with original intentions? How to explain the making and implementation of the SET-Plan?

These research questions are analysed through well-established approaches to EU integration and policymaking, approaches that will emphasize the varying roles of different actors and institutions in why and

---

[3] The package sets three key targets for 2020: 20 per cent cut in GHG emissions (from 1990 levels), 20 per cent of EU energy from renewables, 20 per cent improvement in energy efficiency.

[4] Buchan (2009), Jordan et al. (2010), Oberthür and Pallemaerts (2010), Parker and Karlsson (2010), Birchfield and Duffield (2011), Morata and Sandoval (2012), Boasson and Wettestad (2013), Buchan and Keay (2015), Dupont and Oberthür (2015), Selin and VanDeveer (2015), Torney (2015), Delreux and Happaerts (2016), Dupont (2016), Liefferink and Wurzel (2016), Skjærseth et al. (2016), Wurzel et al. (2017).

[5] A few journal articles have described the contents of the SET-Plan (Soriano and Mulatero 2011), assessed its early achievements in directing R&I funding (Wiesenthal et al. 2009) and indicated its future role in EU energy and climate policy (Ruester et al. 2014).

how the Plan came into being, how it was designed, and how it was later implemented. The Liberal Intergovernmentalism (LI) approach explains outcomes at the EU level mainly in terms of the interests and preferences of the member-states. From this perspective, we would expect the member-states to have had the upper hand in establishing, designing and implementing the SET-Plan in line with their low-carbon R&I priorities. By contrast, the Multilevel Governance (MLG) approach contests the view that member-state governments are in full control of policymaking in Brussels and brings in different explanatory factors and mechanisms in the shaping and implementation of EU policy. From this perspective, we would expect that, in addition to the member-states, the EU institutions and non-state actors at EU level (like industry actors and the research community) will determine the low-carbon priorities in the SET-Plan. However, the SET-Plan does not operate in a vacuum. We also explore various factors external to the Plan, including alignment with EU climate and energy policies and opportunities for EU companies in the international technology market.

The book contributes to the social science literature on EU climate and energy policy by highlighting the importance of *politics* in innovation. The lessons drawn are politically relevant, given the EU's current efforts at reforming its climate and energy policy, including the SET-Plan. The EU's climate, energy and low-carbon technology policies are currently undergoing reform as part of the new Energy Union strategy[6] with new targets and policies set for 2030.[7] Political relevance also extends beyond Europe, given the Paris Agreement's emphasis on spurring wider international cooperation in low-carbon technology innovation.

We apply a theoretically informed case-study approach for analysing the development of the SET-Plan. Our material includes primary data collected in interviews with key actors involved in making and implementing the Plan, and secondary data obtained from document analysis (see Chap. 2).

Chapter 2 presents the analytical framework for answering the research questions. Chapter 3 assesses the development of the SET-Plan. Here we

---

[6] Energy research and innovation have been selected as one of five pillars of the new Energy Union. The other pillars are energy security, a fully integrated internal energy market, improved energy efficiency and emissions reductions.

[7] In 2014, the EU decided to increase the ambitiousness of its climate and energy policy towards 2030 by setting new and stricter targets for GHG reduction (40 per cent by 2030 compared to 1990) and energy-system restructuring (27 per cent share for renewables in EU energy consumption, and 27 per cent improvement in energy efficiency). The targets for renewables and energy efficiency were later increased to 32 per cent and 32.5 per cent, respectively.

present the background for launching the idea of a SET-Plan, the initiation of the Plan and how it was decided by the Commission, the Council and the European Parliament. Further, we examine how the SET-Plan was implemented from 2009, tracking technology priorities in the Plan, funding and organization of low-carbon energy R&I resources in specific programmes and projects. Chapters 4 and 5 provide analyses of why and how the SET-Plan was established and implemented, based on theory-informed expectations introduced in Chap. 2. Chapter 6 sums up the analysis and presents our main conclusions, lessons and prospects for the SET-Plan.

REFERENCES

Birchfield, V. L., & Duffield, J. S. (Eds.). (2011). *Toward a Common European Union Energy Policy: Problems, Progress, and Prospects*. New York: Palgrave Macmillan.

Boasson, E. L., & Wettestad, J. (2013). *EU Climate Policy: Industry, Policy Interaction and External Environment*. Farnham: Ashgate.

Bossink, B. A. G. (2015). Demonstration Projects for Diffusion of Clean Technological Innovation: A Review. *Clean Technologies and Environmental Policy, 17*, 1409–1427.

Buchan, D. (2009). *Energy and Climate Change: Europe at the Crossroads*. Oxford: Oxford University Press.

Buchan, D., & Keay, M. (2015). *Europe's Long Energy Journey: Towards an Energy Union?* Oxford: Oxford University Press for the Oxford Institute of Energy Studies.

Commission. (2007). *A European Strategic Energy Technology Plan (SET-Plan) Full Impact Assessment*. SEC (2007) 1508/2. Brussels: European Commission.

Delreux, T., & Happaerts, S. (2016). *Environmental Policy and Politics in the European Union*. London: Palgrave.

Dupont, C. (2016). *Climate Policy Integration into EU Energy Policy: Progress and Prospects*. Abingdon/New York: Routledge.

Dupont, C., & Oberthür, S. (Eds.). (2015). *Decarbonization in the European Union: Internal Policies and External Strategies*. Basingstoke: Palgrave Macmillan.

IEA. (2015). *Energy Technology Perspectives 2015*. Paris: IEA/OECD.

Jordan, A., Huitema, D., van Asselt, H., Rayner, T., & Berkhout, F. (Eds.). (2010). *Climate Change Policy in the European Union: Confronting the Dilemmas of Mitigation and Adaptation?* Cambridge: Cambridge University Press.

Liefferink, D., & Wurzel, R. K. W. (2016). Environmental Leaders and Pioneers: Agents of Change? *Journal of European Public Policy.* Published online 29 April. https://doi.org/10.1080/13501763.2016.1161657.

Morata, F., & Sandoval, I. (Eds.). (2012). *European Energy Policy: An Environmental Approach.* Cheltenham: Edward Elgar.

Myers, S. (1978). *The Demonstration Project as a Procedure for Accelerating the Application of New Technology.* Washington, DC: Institute of Public Administration.

Oberthür, S., & Pallemaerts, M. (Eds.). (2010). *The New Climate Policies of the European Union.* Brussels: VUB Press.

OECD. (2013). *OECD Frascati Manual* (6th ed.). Updated 11 June 2013. https://stats.oecd.org/glossary/detail.asp?ID=2688

Parker, C. F., & Karlsson, C. (2010). Climate Change and the European Union's Leadership Moment: An Inconvenient Truth? *Journal of Common Market Studies, 48*(4), 923–943.

Ruester, S., Schwenen, S., Finger, M., & Glachant, J.-M. (2014). A Post-2020 EU Energy Technology Policy: Revisiting the Strategic Energy Technology Plan. *Energy Policy, 66,* 209–217.

Selin, H., & VanDeveer, S. D. (2015). *European Union and Environmental Governance.* London: Routledge.

Skjærseth, J. B., Eikeland, P. O., Gulbrandsen, L. H., & Jevnaker, T. (2016). *Linking EU Climate and Energy Policies: Decision-Making, Implementation and Reform.* Cheltenham: Edward Elgar.

Soriano, F. H., & Mulatero, F. (2011). EU Research and Innovation (R&I) in Renewable Energies: The Role of the Strategic Energy Technology Plan. *Energy Policy, 39,* 3582–3590.

Torney, D. (2015). *European Climate Leadership in Question: Policies Toward China and India.* Cambridge, MA: MIT Press.

Wiesenthal, T. G. L., Haegeman, K., & Schwarz, H.-G. (2009). Bottom–Up Estimation of Industrial and Public R&D Investment by Technology in Support of Policy-Making: The Case of Selected Low-Carbon Energy Technologies. *Research Policy, 41,* 116–131.

Wurzel, R. K., Connelly, J. M., & Liefferink, D. (Eds.). (2017). *Still Taking a Lead? The European Union in International Climate Change Politics.* London: Routledge.

Analytical Framework

This chapter outlines the analytical framework that will guide our analysis of the SET-Plan. We combine various strands of theories in order to examine three research questions: (1) How was the SET-Plan established? (2) Have implementation and performance been in line with the original intentions? (3) How can we explain the making and implementation of the SET-Plan?

2.1   Development of the SET-Plan

2.1.1   Making the Plan

The decisions needed to make the SET-Plan concerned policy principles along two dimensions of governance: *who* should govern energy research and innovation policies in the EU, and *how* the development of energy technology should be promoted.

The SET-Plan can first be assessed in terms of how it institutionalized the distribution of competence between the member-states and the EU institutions. A high degree of competence at member-state level would enable national governments to take their diverse individual R&I situations into account (bottom-up/decentralization). By contrast, a high degree of competence at the EU level would enable the EU institutions to harmonize R&I governance across the various member-states (top-down/centralization).

© The Author(s) 2020
P. O. Eikeland, J. B. Skjærseth, *The Politics of Low-Carbon Innovation*, https://doi.org/10.1007/978-3-030-17913-7_2

Secondly, the SET-Plan can be assessed against the policy principle of 'technology neutrality' (Azar and Sandén 2011; Fischer et al. 2012). A technology-neutral policy would combine technology-push and market-pull instruments, so that market agents—not governments—would pick the winners. Technology-push policies would include government support to general R&I infrastructure/capacity-building measures and programmes allowing support for a broad range of emerging and competing energy technologies. Market-pull policies could include carbon pricing and removal of technology-specific subsidies. By contrast, energy technology-specific policies would include narrow public R&I programmes for certain selected technologies combined with technology-specific market-pull policies, such as for renewable energy. Here, the public authorities pick the winners. Between these extremes, other combinations of policies can be imagined, like narrow technology R&I programmes combined with general market-pull instruments (Table 2.1).

A basic expectation is that the intensity and scale of interest mobilization will increase as technology selection deviates from the principle of technology neutrality and as technology governance is centralized. The narrower the priorities, the more technology areas and related actor interests will be left out: they will be expected to mobilize in order to get in. The direction and intensity of mobilization will depend on centralization. Highest mobilization and political conflict can be expected if the EU institutions pick the winners. In this case, member-states will mobilize, as will non-state actors (such as industry and the research community). We would assume that common EU policies will promote a level playing-field and generate less mobilization if the market picks the winners.

Table 2.1   EU energy technology governance: ideal types

	Technology neutrality	
Centralization	Low	High
High	EU institutions pick winners	EU institutions stimulate market to pick winners
Low	Member-state governments pick winners	Member-state governments stimulate market to pick winners

2.1.2 Implementing the Plan

Implementation means to 'carry something into effect' (Weale 1992, p. 43). In our case, implementation relates ultimately to the set of consequences flowing from the SET-Plan according to its goals. The Plan aimed at focusing, strengthening and giving coherence to energy research and innovation in order to accelerate low-carbon technologies (Commission 2007). To achieve these goals, various tasks were envisaged for different actors in the governance system of the SET-Plan.

A first task was to consolidate the choice of priorities (specific technologies and project types) through detailed, coordinated planning of joint R&I programmes and roadmaps, including specification of required resources. A next task was to provide for the higher level of required resources (financial and human) needed to realize the priorities. A third task was coordination among funding sources, to pool resources in joint programmes and demonstration projects. A final task concerned the actual realization of large-scale demonstration projects that could lead to the commercialization of selected technologies.

To assess implementation, we scrutinize the performance of these tasks from 2009. We examine the first task—consolidating selected technologies—by tracking developments in SET-Plan priorities of low-carbon technologies. Was the initial choice of technologies consolidated? Or did the priorities change, with technologies subtracted or added? We then examine the next tasks—providing and coordinating resources—by scrutinizing joint R&I activities and the extent to which funding from private and public R&I budgets (EU and national) was directed towards SET-Plan priorities. Did the selected technologies and large-scale demonstration projects get more funding by strengthening and coordinating funding sources over time? Finally, we examine the extent to which large-scale demonstration projects were realized, to accelerate the deployment of prioritized technologies for the market.

2.2 Explaining Development of the SET-Plan

The SET-Plan envisaged extending EU competence in energy research and innovation policies. It was devised and implemented through the combined political initiatives of the member-states in the Council of the EU and the European Council, the EU institutions (Commission and Parliament), and non-state actors (industry and the research community).

Contextual drivers included climate and related energy market-pull policies evolving within the EU, and international markets for low-carbon technology outside the EU.

2.2.1   The Role of the Member-States

Liberal Intergovernmentalism (LI) places main emphasis on the member-state level, seeing national governments as 'gatekeepers' between national and international politics. It explains EU policy bargaining outcomes as mainly the result of the interests and preferences of these states, leaving scant room for autonomous EU institutions to influence policymaking significantly (Moravcsik 1998, 1999). Basically, 'policy-making in the EU is determined primarily by national governments constrained by political interests nested within autonomous national areas' (Hooghe and Marks 2001, p. 3). According to LI, the principle of national self-determination will prevail. EU-level policy integration will take place if member-state governments prefer more cooperation at the EU level, as emphasized by the New Intergovernmentalism approach (Hodson 2013; Bickerton et al. 2015).

LI leads us to expect member-state R&I interests and priorities to be the key to explaining the making of the SET-Plan. We would expect that the member-states requested the SET-Plan by launching the idea and shaping the Plan's design elements, as energy R&I policies before the SET-Plan had been mainly in the hands of the EU member-states. Any selection of priority technologies would reflect diversity in R&I interests and priorities in the major member-states. We would therefore not expect resources to be concentrated on only a few technologies. Further, we would expect a decentralized governance system controlled by the member-states. Concerning implementation, member-state R&I interests and priorities would be expected to determine the total level of energy R&I resources and alignment between these resources and SET-Plan priorities—also for large-scale demonstration projects.

2.2.2   The Role of EU Institutions and Non-State Actors

Multilevel Governance (MLG) approaches to EU policymaking have been presented as alternatives to state-centred intergovernmentalist approaches (Marks et al. 1996; Fairbrass and Jordan 2004). Variants of MLG all share the assumption that national governments are *not* in full control of policymaking in Brussels, and that European integration has weakened the powers of the state. In our context, that would mean weakened state control over R&I activities. In explaining EU policy, this approach brings in many levels of governance—including the EU, member-state and the subnational levels—thereby blurring any clear separation of domestic and international politics.

Two mechanisms, that 'challenge' state control, are particularly relevant for this study. First, we note the role and influence of the EU institutions acting as organizations and arenas for collective decision making. The Commission is the main 'agenda-setter' and can have an independent influence on policymaking above and beyond its role as agent for national governments (Marks et al. 1996; Sandholtz and Stone Sweet 1998; Skjærseth 2017). This independent role will be conditioned by the Commission's capacity to coordinate internally and with the European Parliament on the initiatives and policy proposals originating in the various units, with their diverging preferences and cultures (Hix 2005; Eikeland 2011; Delreux and Happaerts 2016; Skjærseth et al. 2016).[1] And secondly, there are non-state actors, such as industry and the research community, that seek to influence climate and energy policymaking at the EU level in Brussels, a mechanism that has gained importance as a result of the growing institutional powers of EU institutions (Hix 2005). Combining the two mechanisms, EU institutions may prove particularly influential if transnational policy 'networks' can be formed with industry and the research community, through these networks mustering support, commitment and allegiance to EU-level policies (Marks et al. 1996; Richardson 1996; Sandholtz and Stone Sweet 1998; Bache and George 2006). The development of the SET-Plan is compatible with MLG to the extent that the Plan was shaped also by factors other than member-state interests, including those of EU institutions and non-state actors.

---

[1] Several units of the Commission have roles in policy development related to the SET-Plan. Main responsibility is with DG Energy and DG Research. Other important units include DG Environment/DG Climate Action and DG Mobility and Transport. DG Transport and DG Energy were merged in 2000 under the name DG TREN. In 2010, the two were split again with new acronyms DG ENER and DG MOVE (Mobility and Transport). As of 1 January 2011, the Commission's DG Research officially became DG for Research and Innovation. DG Environment was split in 2010 with a new DG Climate Action responsible for EU climate policies. For simplicity, we will use the terms DG Energy, DG Research and DG Environment/Climate.

Specifically, the MLG approach leads us to expect that that the SET-Plan was initiated and shaped not by the member-states but by the EU institutions and non-state actor networks linked to the EU level. Under the MLG, we further expect the initiation and governance of the Plan to be conditioned by degree of EU institutional unity of preferences. Still, the governance system will be more centralized than expected from LI and controlled by EU-level institutions. Selection of initial technologies will reflect preferences and priorities of the EU institutions. Regarding implementation, we expect opposition from, and mobilization by, member-states and non-state actors that have R&I interests not included as priorities for the Plan. To the extent that EU institutions are unified and allied with non-state actor networks, we expect implementation to reflect EU institutional preferences as to priorities and resources to be provided for low-carbon energy R&I, the alignment between resources and Plan's priorities, and the realization of SET-Plan demonstration projects.

2.2.3   Contextual Factors

The SET-Plan did not evolve in a vacuum, so we explore whether and how two main external factors might have affected its establishment and implementation. First, the Plan aimed at securing the competitiveness and uptake of innovations from European technology firms in international technology markets. We might thus expect decisions on and initial selection of technologies to reflect the scope for creating opportunities for EU technology manufacturers in the international market. That would imply selecting the technologies only if they fill a niche in the international market for technologies that promote or consolidate 'first-mover' advantages.

Second, lessons from innovation studies show that demand-pull instruments play a crucial role in policy mixes, together with technology-push instruments (Mowery et al. 2010; Rogge and Hoffmann 2010; IPCC 2014). Close-to-market low-carbon demonstration projects related to the SET-Plan will be regarded as increasingly risky for private investors and the demonstration programme will appear less attractive if wider market-pull policies are weak or unstable and are expected to remain so (Negro et al. 2012; Åhman et al. 2018). By contrast, strong and stable market-pull policies will reduce the risks of investing in large-scale demonstration plants. The SET-Plan was intended to push technology costs down, while parallel market policies would pull the deployment of low-carbon technologies. We expect implementation of EU climate and energy policies to determine the implementation of SET-Plan priorities, especially concerning realization of large-scale demonstration projects that are closest to the market. Variation in the implementation of EU market deployment policies will affect the realization of large-scale demonstration in different technological areas.

2.3 METHOD

Our main research approach involves a theoretically informed case study. From the theoretical perspectives, we have generated empirically observable expectations to be used as heuristic devices and lenses rather than hypotheses to be rigorously tested. The theories and expectations identify explanatory factors and mechanisms, provide a focus for data collection, and enable us to draw conclusions about the relative merits of alternative explanatory perspectives.

We apply process tracing as the main tool for formulating inferences as to the factors behind the establishment of the SET-Plan and its current state of implementation. Concerning SET-Plan establishment, we identify actors seeking influence over EU SET-Plan governance principles and design. We study the establishment of the SET-Plan as a sequence of policy processes from the idea was launched in 2006; stakeholder responses to the idea in consultations prior to the first general proposal from the Commission in early 2007; and new inputs from the Council, European Council and the Parliament preceding the final Commission proposal of November 2007.

As to implementation, we trace how various stakeholders have engaged from 2009 in performing key tasks as prescribed by the SET-Plan: to consolidate technology priorities, increase the level of R&I resources and coordinate these in joint programmes and large-scale demonstration projects to realize innovations in energy technology.

Our study applies qualitative data for analysing the establishment of the SET-Plan and a mixture of qualitative and quantitative data for analysing its implementation. Triangulation is used to cross-check data from the various sources. For primary data we draw on a series of 21 semi-structured interviews conducted with persons engaged in the establishment and implementation of the SET-Plan (see list of interviews): staff in the Commission (and its Joint Research Centre); national government representatives on the SET-Plan; and representatives of national research institutions, EU associations of research institutes and industry. Secondary data include public policy papers of the Commission (communications and impact assessments, consultations and proposals for the SET-Plan); implementation reports and statistics from the Commission and its Joint Research Centre; minutes from meetings of the Council and the European Council; and Parliamentary motions, reports and minutes from plenary debates.

2.4   SUMMARY

This chapter has developed an analytical framework for assessing and explaining the SET-Plan. We distinguish between two policy phases—the making and the implementation of the Plan—and will explain these phases by two theoretical approaches: Liberal Intergovernmentalism and Multilevel Governance. Contextual factors are included to explore the relationships between the SET-Plan and EU climate and energy policies and international technology markets (Table 2.2).

Table 2.2   Theories, expectations and policy phases in SET-Plan development

	Making the Plan	*Implementing the Plan*
LI	Member-states requested and shaped the SET-Plan. The governance system will be decentralized. Any deviation from technology neutrality will reflect diversity in R&I priorities among major member-states. Priorities will not be concentrated on only a few technology areas	Member-state R&I interests and priorities determine the level of R&I resources, alignment between resources and SET-Plan priorities and the realization of demonstration projects
MLG	The EU institutions took initiative independently and shaped the Plan based on existing non-state actor networks. The governance system will be more centralized and controlled by EU institutions. R&I priorities will reflect EU institutional preferences	If unified, EU institutions will overcome opposition from those with different priorities and determine the level of R&I resources, alignment between resources and SET-Plan priorities and the realization of demonstration projects
Contextual factors	Technologies will be selected only if they fill a niche in the international market that promotes 'first-mover' advantages	Implementation of EU climate and energy policies aimed at pulling market deployment will determine variation in the realization of large-scale demonstration projects for different technologies

REFERENCES

Åhman, M., Skjærseth, J. B., & Eikeland, P. O. (2018). Demonstrating Climate Mitigation Technologies: An Early Assessment of the NER 300 Programme. *Energy Policy, 117*, 100–107.

Azar, C., & Sandén, B. A. (2011). The Elusive Quest for Technology-Neutral Policies. *Environmental Innovation and Societal Transitions, 1*(1), 135–139.

Bache, I., & George, S. (2006). *Politics in the European Union* (2nd ed.). Oxford: Oxford University Press.

Bickerton, C. J., Hodson, D., & Puetter, U. (2015). The New Intergovernmentalism: European Integration in the Post-Maastricht Era. *Journal of Common Market Studies, 53*(4), 703–722.

Commission. (2007). *A European Strategic Energy Technology Plan (SET-Plan): 'Towards a low carbon future'.* COM (2007) 723 Final. Brussels: European Commission.

Delreux, T., & Happaerts, S. (2016). *Environmental Policy and Politics in the European Union.* London: Palgrave.

Eikeland, P. O. (2011). The Third Internal Energy Market Package: New Power Relations Among Member States, EU Institutions and Non-state Actors? *Journal of Common Market Studies, 49*(2), 243–263.

Fairbrass, J., & Jordan, A. (2004). Multi-level Governance and Environmental Policy. In I. Bache & M. Flinders (Eds.), *Multi-level Governance* (pp. 147–164). Oxford: Oxford University Press.

Fischer, C., Torvanger, A., Shrivastava, M. K., Sterner, T., & Stigson, P. (2012). How Should Support for Climate-Friendly Technologies Be Designed? *Ambio, 41*(Suppl 1), 33–45.

Hix, S. (2005). *The Political System of the European Union.* New York: Palgrave.

Hodson, D. (2013). The Little Engine That Wouldn't: Supranational Entrepreneurship and the Barroso Commission. *Journal of European Integration, 35*(3), 301–314.

Hooghe, L., & Marks, G. (2001). *Multi-level Governance and European Integration.* Lanham: Rowman & Littlefield.

IPCC. (2014). *Climate Change 2014: Synthesis Report. Contribution of Working Groups I, II and III to the Fifth Assessment Report of the Intergovernmental Panel on Climate Change.* Geneva: IPCC.

Marks, G., Hooghe, L., & Blank, K. (1996). European Integration from the 1980s: State-Centric vs. Multi-level Governance. *Journal of Common Market Studies, 34*(3), 341–378.

Moravcsik, A. (1998). *The Choice for Europe: Social Purpose and State Power from Messina to Maastricht.* London: Routledge.

Moravcsik, A. (1999). A New Statecraft? Supranational Entrepreneurs and International Cooperation. *International Organization, 53*(2), 267–396.

Mowery, D. C., Nelson, R. R., & Martin, B. R. (2010). Technology Policy and Global Warming: Why New Policy Models Are Needed (Or Why Putting New Wine in Old Bottles Won't Work). *Research Policy, 39*(8), 1011–1023.

Negro, S. O., Alkemade, F., & Hekkert, M. P. (2012). Why Does Renewable Energy Diffuse So Slowly? A Review of Innovation System Problems. *Renewable and Sustainable Energy Reviews, 16*(6), 3836–3846.

Richardson, J. J. (1996). Policy-Making in the EU: Interests, Ideas and Garbage Cans of Primeval Soup. In J. Richardson (Ed.), *European Union: Power and Policy-Making* (pp. 3–23). London: Routledge.

Rogge, K. S., & Hoffmann, V. H. (2010). The Impact of the EU ETS on the Sectoral Innovation System for Power Generation Technologies—Findings for Germany. *Energy Policy, 38*(12), 7639–7652.

Sandholtz, W., & Stone Sweet, A. (1998). *European Integration and Supranational Governance.* Oxford: Oxford University Press.

Skjærseth, J. B. (2017). The European Commission's Shifting Climate Leadership. *Global Environmental Politics, 17*(2), 84–104.

Skjærseth, J. B., Eikeland, P. O., Gulbrandsen, L. H., & Jevnaker, T. (2016). *Linking EU Climate and Energy Policies: Decision-Making, Implementation and Reform.* Cheltenham: Edward Elgar.

Weale, A. (1992). Implementation Failure: A Suitable Case for Review? In E. Lykke (Ed.), *Achieving Environmental Goals: The Concept and Practice of Environmental Performance Review.* London: Belhaven Press.

CHAPTER 3

Development of the SET-Plan

This chapter begins with an assessment of the making of the SET-Plan, focusing on governance and the criteria for selecting low-carbon technology priorities. We then turn to whether and how the SET-Plan has been implemented in line with its goals of focusing, strengthening and giving coherence to EU energy research and innovation.

3.1   MAKING THE PLAN

3.1.1   Background

Energy Research and Development (R&D) has been part of EU[1] policy ever since the European Coal and Steel Community Treaty (ECSC) of 1951 entasked the new High Authority (predecessor of the Commission) with coordinating technical and economic research on coal (European Parliament 2016).[2] Nuclear energy research was at the heart of the second founding Treaty: the EURATOM Treaty of 1957 established the Joint

---

[1] For simplicity, we apply the term EU throughout the book, also for the period before the EU was formed.

[2] When the ECSC Treaty expired in 2002, revenues were transferred to the EU in a new Research for Coal and Steel Fund (RFCS). Some €50 million per year of EU funding was made available to universities, research centres and private companies to fund projects (Commission 2015c).

© The Author(s) 2020
P. O. Eikeland, J. B. Skjærseth, *The Politics of Low-Carbon Innovation*, https://doi.org/10.1007/978-3-030-17913-7_3

Nuclear Research Centre as a new EU-level institution with research sites and laboratories in several EU countries.[3]

In the 1970s, the EU diversified its energy research to include a broader portfolio of indigenous energy sources and technologies. The Joint Nuclear Research Centre was renamed the Joint Research Centre (JRC) to reflect this diversification that included a new laboratory for testing solar photovoltaic (PV) devices (ESTI 2016). The background was the international oil crisis that constrained oil imports, sent international oil prices skyrocketing and hampered economic growth.

The scale and scope of EU energy R&D were extended in the 1980s when the Council adopted the first Framework Programme on Scientific Research and Development (FP1, 1984–87). Nearly half the budget was allocated to energy research, with more or less equal shares for nuclear energy and renewable energy/energy-efficiency projects (Council 1983a; Andrée 2009). Similar research priorities were determined for the JRC, which had evolved as a major EU-level research institution (Council 1983b).[4] The EEC Treaty was reformed in 1986, providing a stronger legal basis for Community research policies through the addition of 'Research and Technological Development' as a separate title.

Despite this stronger legal basis, the EU energy R&D budget brought a sharp cut in FP2 (1987–1991), reflecting cuts also in member-states and industry (Andrée 2009). The quest for new energy efficiency and renewable energy technology to substitute oil imports lessened as international oil prices plunged in 1986: security of supply and competitiveness concerns became less imminent. Non-nuclear energy R&D budgets were kept relatively stable at a low level in the next three successive Framework Programmes (FPs 3–5, 1990–2002) (van der Veen et al. 2014, p. vi).

Low priority for energy research concurred with unsuccessful political responses to the emerging issue of climate change (Skjærseth 1993, 1994). The Commission first addressed the issue in its research policy in 1985. Three years later, the first communication from the Commission to the Council was presented (Commission 1988), urging the industrial countries to commit themselves firmly to stabilizing CO_2 emissions by the year 2000.

A joint Council of energy and environmental ministers in 1990 followed up and agreed on stabilizing EU CO_2 emissions by that year. The

---

[3] Belgium, Germany, Italy, the Netherlands and Spain.
[4] At that point, JRC employed around 2200 staff (Council 1983b).

Commission next proposed a comprehensive package of climate and energy policies, to lay the foundations for a lead role at the 1992 UN Conference on Environment and Development in Rio de Janeiro. This package included both market-pull and technology-push instruments, with an EU-level carbon/energy tax, a funding programme for demonstrating non-nuclear energy technologies (the THERMIE Programme) and support to accelerate market deployment of energy efficiency and renewable energy technologies (the SAVE and ALTENER programmes). However, the highly contested tax proposal was blocked by the Council, and the proposed research and innovation programmes for low-carbon technologies were adopted only after cuts to proposed budgets. Climate and energy policies largely failed to provide a forceful market stimulus to low-carbon energy solutions (Skjærseth 1994).

In the wake of the 1997 Kyoto Protocol, the EU made new efforts to adopt climate and energy market-pull policies. Agreement was reached on a series of new policies that included the 2003 EU Emissions Trading Directive (EU ETS), which placed a price on carbon. New directives were adopted to stimulate demand for renewable electricity, biofuels and energy efficiency in buildings. However, these policies were neither binding nor particularly ambitious (Skjærseth et al. 2016). Joint energy policies on renewables and energy efficiency were weakened by the principle that each member-state should have full sovereignty over its own energy mix. Moreover, whether the EU should take an international climate leadership-by-example role was now seriously contested within the Commission and by some member-states (Skjærseth et al. 2016).

As the twentieth century drew to a close, rising oil prices and growing import dependencies revived deep concerns for EU energy security and affordability (Commission 2000a).[5] New challenges concerned EU industrial competitiveness vis-à-vis the United States and Japan linked to weaker research and innovation policies, in general and for the energy sector specifically. Compared to the United States and Japan, the EU was spending fewer resources on energy R&D, and a lower share of resources went to market-near demonstration of low-carbon energy technologies. EU-level energy programmes lacked strategic focus: resources were scattered across many topics and programmes for energy R&D and demonstration projects were poorly coordinated (Wejnen et al. 2000).

---

[5] Dependence on external energy sources was expected to reach 70 per cent in 2030.

In response to these challenges, the Commission (DG Energy) adopted a Green Paper sketching a plan for how to increase the share of clean, durable and renewable energy use through directing community support towards market deployment and large-scale demonstration projects (Commission 2000a).[6] The first seeds of a future EU strategic energy technology plan were thus sown at the turn of the century.

In parallel, DG Research proposed, and the Lisbon European Council adopted, the European Research Area (ERA), aimed at making the EU the most competitive and dynamic knowledge-based economy in the world by 2010 (the 'Lisbon Strategy'). Better coordination of resources at EU, national and industry levels was called for, as well as a shifting of resources towards innovation and market diffusion of technologies (European Council 2000). Rectifying the fragmentation and compartmentalization of national research generally was recognized as crucial for success in transiting to an internationally competitive knowledge-based economy (Commission 2000b). This new strategy also included mitigation of climate change as an objective for long-term technology development and competitiveness (Skjærseth et al. 2016).

To govern implementation of ERA, the European Council decided to apply a new Open Method of Coordination (OMC), which would include networking national and joint research programmes on a voluntary basis (European Council 2000). The governance model would give the member-states considerable control over EU R&I policy (European Council 2000; Armstrong et al. 2008).[7] According to further implementation guidelines proposed by the Commission, EU-level research efforts should be complementary to member-state efforts, focused on a more limited number of priorities subject to political choice and based on objective assessment criteria to create European added value (Commission 2000c).[8] Network instruments would be developed by the Commission

---

[6] Shifting existing support towards markets would create economies of scale, eventually making low-carbon technologies cheaper (Commission 2000a).

[7] The emergence of the OMC generally reflected member-state concerns about EU institutions extending their competence and powers in areas where the subsidiarity principle was meant to apply (Armstrong et al. 2008).

[8] Criteria applied in past FP decisions included 'cost and scale of research above and beyond the possibilities of a single country, and the need to assemble a critical mass of financial and human resources', and 'links with EU priorities and interests and with Community legislation and policies' (Commission 2000c).

for coordination at the EU level, including technology platforms for clustering industrial research activities.

High political attention was maintained on closing the innovation gap between European industries and the main competitors. In 2002, EU leaders pledged to increase national spending on research and innovation from 1.9 per cent to 3 per cent of GDP—of which two-thirds should come from the private sector (European Council 2002). In March 2003, the European Council called for concrete action on the part of the member-states to promote business investments in R&I (European Council 2003). The Commission followed up with an EU action plan proposing increased participation of industry and other stakeholders in the determination of priorities for public research, and more structured partnerships between industry, universities and research organizations to promote academic research as well as technology-based innovation (Commission 2003).

Further implementation of the ERA brought about notable organizational changes affecting how resources were prioritized and used at the EU level. FP6, created with far more input from 'below', entailed increased spending for networking and partnerships among research communities, member-states and industry (Commission 2007c; Andrée 2009; van der Veen et al. 2014). New networks included European Technology Platforms (ETPs), industry-led forums for identifying strategic research agendas and for prioritizing and mobilizing research resources in various areas of technology, including energy technologies (Commission 2004). The first energy-sector ETPs established in 2004 by the Commission's DG Research were on 'Fuel Cells and Hydrogen' and 'Photovoltaics' (Commission 2004). ETP initiatives that could benefit mainly from community funding were selected as Joint Technology Initiatives—a new public–private partnership funding mechanism (Commission 2004). Selection of Technology Initiatives would be made in close collaboration with EU member-state authorities. To stimulate networking among managers of national research programmes, the ERA-NET Scheme was established. The scheme would cover coordination costs related to creating joint research programmes, also in the field of energy.[9] A new European Institute of Innovation and Technology was to foster partnerships between businesses and research

---

[9] Under FP7, a new ERA-NET Plus mechanism would extend funding for joint member state projects.

institutes: Knowledge and Innovation Communities (KICs), which would cover also energy research and innovation from 2009.[10]

Despite the Lisbon Strategy and ERA initiatives, energy technology research and innovation efforts at EU and national levels still had to struggle with greatly constrained budgets and poor coordination (Commission 2007d, p. 7). The new FP6 (2002–2006), decided by the Parliament and Council in 2002, had a lower budget for low-carbon energy research than FP5, although it was adjusted upwards after pressure from the European Parliament (European Parliament and the Council 2002). Funding of energy research continued its long-term declining trend also at national and industry levels, with particularly weak budgets in the new Central and East European (CEECs) member-states from 2004.

The politically promised reorientation of resources towards more support of industrial innovation and market deployment had little immediate impact. Great variation remained among the member-states in resources allocated specifically for industrial demonstration and deployment of energy technologies; also, EU-level budgets for low-carbon energy demonstration projects remained short in supply (Commission 2007d; European Parliament 2008a).[11]

In summary, the period before the SET-Plan was initiated was characterized by declining public and private funding of energy technology research and innovation in the EU, acknowledged as a major problem since the challenges of climate change, security of supply and industrial competitiveness required faster cost-reduction of new technologies. Policies to push low-carbon energy technology development lacked strategic focus, and EU market-pull policies were weak. Energy research programmes at national and EU levels were heterogeneous and fragmented.

---

[10] The European Institute of Innovation and Technology, proposed by Commission President Barroso in 2005, was modelled on the Massachusetts Institute of Technology (MIT) in the United States. Meeting heavy opposition from existing universities and member-state governments, the idea was relaunched as mainly a coordination unit for new energy research and innovation networks called Knowledge and Innovation Communities (Gornitzka and Metz 2014). These also included climate and energy research and innovation networks from 2009.

[11] A notable exception was the new Intelligent Energy for Europe Programme adopted in 2003, focusing on removal of non-technological barriers to deployment of new sustainable technologies. This programme sponsors the involvement of a wider group of societal actors, cities, field experts and urban planners for diffusion of new innovative social models for technology deployment (Turmes 2017, p. 19).

However, the seeds of a more strategic joint EU approach to low-carbon energy technology innovation closer to the market had been sown by the Commission in a 2000 Green Paper and by the various networking initiatives and public–private partnerships undertaken by DG Research for implementation of the ERA in the energy sector.

3.1.2 Initiation

At their Hampton Court Summit in autumn 2005, EU leaders signalled that they might accept extended EU competence in energy policy cooperation (Euractiv 2005). The Commission responded in March 2006 by adopting a Green Paper with suggestions and options for a new comprehensive European energy policy (Commission 2006a). International events related to energy security and climate change contributed to the policy momentum: gas supplies to Europe were interrupted by a dispute between Ukrainian and Russian companies; international oil prices continued their upward trend; and public attention to climate change was rising (Skjærseth et al. 2016).

The Green Paper prepared by DG Energy proposed that, in order to deal with the concerns of climate change, energy security and competitiveness, EU energy-policy decision makers should agree on a new target for the energy mix, with higher shares of low-carbon energy sources (Eikeland 2012). As an integral part of the new energy strategy, the Commission proposed a new EU Strategic Energy Technology Plan that would focus EU R&I efforts at EU and national levels towards achieving energy policy goals and targets. Such close coordination between sector and technology policies was something new in Brussels policymaking (Interview D).

The main purpose of such a plan would be to: 'accelerate research in promising energy technologies … [and] create the conditions to bring such technologies efficiently and effectively to the EU and the world markets' (Commission 2006a, p. 13). The proposal thus responded to previously identified needs for better coordination and concentration of resources around strategic priorities to facilitate innovation and deployment.

However, which technologies should qualify as 'promising' for accelerated EU innovation efforts, and how these should be selected, were not elaborated in the Green Paper. Some general formulations indicated that the Plan should build on the experiences of the seven European Technology Platforms established to identify strategic research agendas and deployment strategies in energy technology (Commission 2006a, p. 31):

- Fuel Cells & Hydrogen
- Solar Photovoltaics (PV)
- Solar Thermal
- Electricity Network Technology/smart grids
- Biofuels
- Zero-emissions fossil-fuel power/CCS
- Wind

The Green Paper emphasized that R&D and demonstration of carbon capture and storage (CCS) and clean-coal technologies would be particularly important for countries choosing coal as a secure and abundant energy source (Commission 2006a).[12] Beyond the ETP-focused technology areas, the Green Paper stated that also a broader focus should be contemplated for the Plan, such as research on high energy-use sectors like housing and transport. Nuclear fusion and fission technology projects were used as illustrations of already well-established and concerted EU R&I action.

In March 2006, the European Council endorsed the development of a new coherent and integrated EU energy policy, reaffirming political signals given at Hampton Court in 2005 (European Council 2006). Also endorsed was the principle that policies should be underpinned by a common target for a minimum share of low-carbon energy sources in the EU energy mix.[13] On the specific proposal for a SET-Plan, the European Council was silent, but consented more vaguely to giving energy and low-carbon technologies higher priority in national and community R&D budgets.

To advance the specific SET-Plan idea, a handful of lower-level Commission staff in DG Research and DG Energy who managed the bulk of energy research and demonstration funds at the EU level took work further (see Chap. 4). They argued their case in Commission interservice consultations and got backing from higher decision-making levels to draft the first communication on the Plan (Interviews D and N).

---

[12] The EU energy mix (the proportion of various energy sources in EU primary energy consumption) in 2005 was as follows: coal and lignite (18.1%), natural gas (25.1%), oil (33.8%), nuclear (15%), renewable energy (7.1%), (European Environment Agency 2017). There was great variation among the member states.

[13] The EU leaders agreed to consider raising existing renewable energy and biofuels targets for 2010 (12% and 5.75%) to 15 per cent and 8 per cent by 2015, and to 'bear in mind' the 20 per cent Commission-estimated energy efficiency potential by 2020.

A series of consultations assisted Commission's work on transforming the Green Paper into more concrete proposals for new energy policies. These consultations evoked responses from member-state governments, EU institutions, the research community and industry (Eikeland 2012). Respondents broadly agreed that a strategic technology plan should be included in the portfolio of new EU energy policies but differed widely as to which energy technologies should be promoted.

A questionnaire directed at organizations and citizens revealed the following ranking: (Commission 2006b).[14]

Solar	67.0%
Wind	61.7%
Second-generation biomass	55.8%
Tidal and wave	44.5%
Fuel cells and hydrogen	44.1%
Smart electricity networks	41.0%
CO_2 Capture and Storage (CCS)	24.3%
Disposal of nuclear waste	20.6%
Clean Coal (non-CCS)	12.3%
Enhanced Oil Recovery	6.5%
Other[15]	10.9%

Among the member-state governments submitting written responses, CCS and Clean Coal were accorded high priority by Germany and France, and clean-coal technologies by the Czech Republic and Poland. Nevertheless, the Commission concluded that the consultations had revealed broad support to 'sustainability' as overarching goal for new energy policies, and for the new strategic energy technology plan to give priority to low-carbon energy technologies (Commission 2006b, p. 3; Skjærseth et al. 2016).[16]

On 10 January 2007, the Commission adopted synchronized communications for the EU's future climate and energy policies, prepared by DG Environment and DG Energy, respectively (Commission 2007a, b). A

[14] Most of the responses came from Germany, the United Kingdom and France; the CEECs were poorly represented.

[15] This included energy-saving solutions, optimization of gas-fired power stations, nuclear power, large-scale storage batteries and geothermal energy (Commission 2006b).

[16] This was framed by the Commission as more support given to 'sustainability' based on an EU target for renewable energy than 'security of supply' and 'competitiveness' (Eikeland 2012).

third communication, co-drafted by DG Energy and DG Research, brought forward the SET-Plan idea (Commission 2007c). Central in all three policy documents were what became known as the 20–20–20 targets: to cut GHG emissions by 20 per cent (from 1990 levels) and increase the share of renewables and energy efficiency by 20 per cent, by 2020. All underscored *synergies* between climate, energy and low-carbon technology policies. The SET-Plan would serve as the key low-carbon technology-push instrument. New EU climate and energy legislation for achieving the 2020 targets would serve as market-pull instruments. Policies to stimulate renewables, energy efficiency and energy technology development were to create new green growth and jobs, and thus be compatible with the Lisbon Agenda.

International events continued to strengthen the momentum for new climate and energy policies. The IPCC's 2007 Assessment Report attracted exceptional attention, with its findings that 'warming of the climate system is unequivocal' and would continue unless changes were made to climate-change mitigation policies (IPCC 2007, p. 72). Support for EU climate action was also confirmed by Eurobarometer surveys that showed widespread worries about the risk of climate change and perceptions that industry, citizens, national governments and the EU should do more to fight the problem.[17] Also energy security continued to climb up the EU agenda. Since the mid-1990s, energy-import dependency had been on the rise, reaching 54 per cent in 2006 at an estimated cost of €350 billion, or around €700 per year for every EU citizen (Commission 2008a, p. 2).[18]

The January 2007 communication outlined some broad principles for the SET-Plan. First, it should embrace: 'all aspects of technological innovation, as well as the policy framework required to encourage business and the financial community to deliver and support low carbon technologies' (Commission 2007c, p. 8). Second, it should be inclusive in terms of participation. All relevant actors would be needed to make and implement the Plan: industry, the research community, the financial community, public bodies, users, civil society, citizens and unions. Third, the strategic element would primarily involve identifying and selecting specific low-carbon

---

[17] Three-quarters of those surveyed confirmed that they perceived the risk of climate change very seriously (Eurobarometer 2008).

[18] By 2012, 53 per cent of EU energy consumption was still linked to imports: 88 per cent of crude oil, 66 per cent of natural gas, 42 per cent of coal and 95 per cent of uranium (Commission 2014c).

technologies: 'for which it is essential that the European Union as a whole finds a more powerful way of mobilizing resources … to accelerate development and deployment' (Commission 2007c, p. 8).

The set of criteria on which to base this selection was not made fully clear. However, the Commission stated that initiatives should be large scale, and thus beyond the capacity of any single country—as exemplified by 'smart grids', biorefineries and CCS that would require new cross-border infrastructure. Beyond this, and the reiteration that the strategic agendas of the ETPs could be important, the communication was vague on specific technological priorities, merely stating that the Plan should include technologies at different levels of maturity and avoid 'being perceived as a European level picking winners' (Commission 2007c, p. 8). Still, this seems to be what the Commission had in mind by: 'ensuring that the *right* portfolio of technologies is brought forward to the Member States to pick and choose' (Commission 2007c, p. 8, emphasis added). Apparently, this 'right' portfolio was to be brought forward by the Commission.

Ten transport and 16 electricity/heat-conversion technologies at varying stages en route towards widespread deployment, identified by FP6 Advisory Group on Energy,[19] were listed as reference points, together with vision statements on market penetration prospects from the seven energy ETPs established from 2004 (Commission 2007c, pp. 11–12). Not included were energy-efficiency solutions that also had been broadly assessed by the FP Advisory Group.

CCS technologies and policies stood out as a strong priority candidate, the focus of a separate communication accompanying the policy package that formulated the aim of emissions-free fossil-fuel power plants after 2020 (Commission 2006b). CCS could also have a political role by mobilizing accept and support for EU-level climate and energy policies from the fossil-fuel industry and the heavily coal-dependent new Central and Eastern European member-states, and would allow continued use of indigenous coal resources, thus providing a solution to both energy-security and climate-policy concerns. Many of these CEECs saw security of supply as their greatest energy-policy concern, with Poland—the most coal-dependent EU country—outspoken in advocating the use of domestic coal resources (Eikeland 2012; Skjærseth 2014).

---

[19] The key task of the Advisory Group on Energy (AGE) is to contribute to the implementation of the energy part of the EU FPs and EURATOM Programme by advising the Commission services (Commission 2018c).

In March 2007, the European Council welcomed the key elements of new climate, energy and technology policies and adopted the 2020 targets. The EU leaders agreed that a 20 per cent GHG emissions reduction (compared to 1990) and a 20 per cent share for renewables in total energy consumption (including a 10 per cent target for share of biofuels in transport) should be binding. Also adopted was a non-binding commitment to reduce EU energy consumption by 20 per cent through energy-efficiency investments.

Concerning energy technologies, the European Council welcomed: 'the Commission's intention to table a European Strategic Energy Technology Plan during 2007 for consideration at the latest by the 2008 Spring European Council' (European Council 2007, p. 22). Member-states and the Commission were urged to strengthen research on CCS; the European Council welcomed the Commission's intention to stimulate the construction and operation by 2015 of up to 12 pilot plants and, if possible, demonstration with new fossil-fuel power plants by 2020. However, the European Council was less clear on continued promotion of research in nuclear power and indicated that R&D should focus on nuclear waste management.

Up until summer 2007, the Commission organized comprehensive consultations on the SET-Plan communication, conducting 18 hearings and workshops on what the Plan should prioritize, asking stakeholders to assess 15 possible areas of technology (Commission 2007d, p. 5).[20] These included energy research on fossil fuels and energy efficiency in buildings.

An additional online questionnaire received 604 responses from individuals and organizations (Commission 2007d). The key messages were summarized by the Commission: technologies to lower CO_2 emissions, increase the use of renewable energy and improve energy efficiency were generally seen as important. More than 90 per cent of respondents agreed that EU-level action would add value for promoting energy technologies at different stages of development; 89 per cent recognized the need for a stronger focus on market uptake of low-carbon energy. The SET-Plan idea was widely welcomed, but differences remained as to which low-carbon

[20] Wind energy, biofuels, photovoltaics, electricity networks, hydrogen and fuel cells, zero-emission fossil-fuel power plants, hydrogen and renewable energy, nuclear fusion, nuclear fission, concentrated solar power, solar thermal heating, cogeneration of heat and power, energy efficiency in buildings, geothermal energy, oil and gas, basic sciences and international cooperation.

technologies should be prioritized for EU efforts, and according to which criteria.

In November 2007, the Commission formally proposed the SET-Plan (Commission 2007e). The aim was to correct unique weaknesses trapping low-carbon technologies in the 'valley of death', hindering them from progressing to uptake in the market.[21] Weaknesses included the scale of investments needed, technological and regulatory inertia and lock-in, the lack of natural market interest and short-term business benefits (Commission 2007e, p. 3).

Concerted EU intervention would be needed, to fill the gap in attaining the 2020 and 2050 targets and to take the global lead in energy technologies.[22] It was acknowledged that some technology challenges would require large-scale investment and entail risks which 'cannot be met by the market, by Member States acting individually or by the current model of European collaborative research' (Commission 2007e, p. 6).

The Commission called for a collective endeavour on the part of the private sector, the research community, the member-states and at the Community level to '*focus, strengthen* and *give coherence* to the overall effort in Europe, with the objective of accelerating innovation in cutting edge European low carbon technologies. In doing so, it will facilitate the achievement of the 2020 targets and the 2050 vision' (Commission 2007e, p. 9, emphasis added)

The SET-Plan was intended to deliver four concrete results: (1) new strategic planning, (2) more effective implementation, (3) increase in resources, (4) new and reinforced approach to international cooperation.[23]

To ensure delivery of these results, a new governance system was proposed. This included new EU-level arenas for implementing the Plan among member-states; industrial companies, research institutions and the European Commission, including the JRC. A central role was envisaged for the member-states in a new Commission-chaired Steering Group, with high-level government representatives mandated to conceive joint actions, make resources available through coordination of national programmes, and monitor and review the progress of the Plan. A European energy

---

[21] A phenomenon described in the innovation literature, indicating that many research ideas fail to progress to commercialization in the market (e.g. see Markham et al. 2010).

[22] The main competitors identified were the United States and Japan, and emerging economies such as China, India and Brazil (Commission 2007e).

[23] A reinforced international energy technology cooperation strategy was aimed at, but no new specific actions were proposed.

technology information system (SETIS) was proposed for reporting, monitoring and review—to be developed and administered by the JRC.

Implementation would first be ensured by European Industrial Initiatives (EIIs). These were intended as arenas for mobilizing industry engagement in research and demonstration and would be led by industry—in contrast to the ETPs and the KICs, which had evolved as more academically focused arenas (CEPS 2011, p. 15). Industrial initiatives would be implemented in various ways, from public–private partnerships to coalitions of interested member-states, depending on the nature and needs of the sectors and technologies involved. The aim was, starting from 2008, to mobilize a critical mass of activities and actors for research and innovation for six priority European initiatives:

- Wind: large-scale demonstration for on- and offshore turbines and their adoption in the energy system
- Solar: large-scale demonstration of photovoltaics and concentrated solar power
- Bioenergy: 'next generation' biofuels within the context of overall bioenergy use
- CCS: prove the viability of zero-emissions fossil-fuel power plants at industrial scale
- Electricity grid: development of smart electricity system, including storage
- Nuclear fission: development of sustainable new nuclear reactor designs (Generation IV reactor technologies).

A central driver of the SET-Plan had been the wish to strengthen European businesses, as signified by the Commission's linking selected technologies to the industrial initiatives. The list of selected prioritized technology areas shows that the SET-Plan would give main priority to low-carbon electricity generation. The selected priorities corresponded with agendas of the existing European Technology Platforms, with two exceptions: nuclear fission was included, and fuel cell and hydrogen excluded.[24] Excluded were also several technology areas that had been considered earlier, such as oil and gas, clean-coal, geothermal, ocean and energy-efficiency technologies. It was not fully clear what selection criteria the Commission had

---

[24] It was already planned to resource Fuel Cells & Hydrogen at EU level through a Joint Technology Initiative funded directly under FP7.

applied for the EIIs. The accompanying Impact Assessment highlighted various criteria: EU added value/additionality; the willingness of actors to join forces (member-states and/or industry); potential market penetration of the technology under various time horizons; potential contribution to CO_2 reduction; security of energy supply; and competitiveness (Commission 2007d, p. 17).

As an additional implementation instrument, the Commission proposed that universities and research institutes form a European Energy Research Alliance (EERA) to coordinate and conceive joint research programmes and align with the EIIs—an energy-specific version of the European Research Area.

The SET-Plan and its governance structures would be established under an appropriate legal basis, *or* under the Open Method of Coordination based on the European Research Area (Commission 2007d). The Commission aimed at doubling financial resources to the six priorities. A specific communication on funding was promised at the end of 2008. Timing of the SET-Plan was not optimal here, however: the EU's main R&I funding programme, FP7 for the period 2007–2013, had already been adopted and its main energy priorities decided.

The governance system was based on the Commission's assessment of four alternative policy options (Commission 2007d). These differed in designation of control to national governments and EU institutions over selection of priorities to the Plan and coordination of resources for implementing the Plan. The selected 'Strategic Coordination' option envisaged a governance system with shared competence between member-states and the EU level, rebalanced with more top-down centralized control—greater EU coordination of implementation resources based on priorities set at the EU level. This option was viewed as the politically most feasible solution at the time.

In essence, then, the proposed SET-Plan included six selected low-carbon energy technology areas as priority for coordinated efforts; and a governance system for implementation with a Steering Group; a system for reporting, monitoring, and review; European Industrial Initiatives to focus on industry large-scale demonstration projects; and a European Energy Research Alliance to coordinate national research efforts and align these with efforts of industry within the EIIs. The proposal had discarded a market-driven bottom-up governance option in line with the principle of technology neutrality, aiming instead at a 'mixed' system where competence for picking the winners would be delegated to the member-states, based on the Commission's initial priorities. However, the specific balance of decision-making power between these levels was not entirely clear. Neither were the criteria applied for selecting specific technology areas as EIIs fully clear, but existing ETPs, 'goal steering' and political feasibility had all been important.

3.1.3   Decision Making

In February 2008, the Council gave its general backing to the SET-Plan and the six priority EIIs (Council 2008)—adding, however, that these priorities should have no implications for the provision of financial means. The Council also emphasized that advances should be sought in a broad portfolio of energy technology fields, with energy efficiency and end-use technologies as essential parts of the Plan. The Steering Group was endorsed as a key governance mechanism to be controlled by the member-states with no mention of the Commission's role here. The Council made it clear that the SET-Plan should allow member-states to pursue R&I in line with their own national situations and preferences, and their prerogative to decide on their own energy mixes (Council 2008, p. 2). These restrictions undermined the Commission's intention of strengthening EU-level competences in energy R&I policies. They also meant that the Hampton Court signals regarding extended EU coordination in energy policy were apparently not to apply to the policy field of energy R&I.

In March 2008, also the EU Heads of States followed up with general backing of the SET-Plan (European Council 2008). From March to July 2008, the SET-Plan proposal was debated in the European Parliament (EP), ending in a resolution (European Parliament 2008b). The EP supported the six EIIs—but, like the Council, argued for a broader portfolio of technologies. Also requested were additional criteria for selecting technologies in connection with the Plan: a clearer focus on the contribution to 2020 targets and consideration to be given to 'life-cycle impacts' and trade-offs between low-carbon energy sources and nature conservation. The governance system did not cause much debate—the concept of the Steering Group was supported; likewise, the proposed SETIS information system, the EERAs, and the EIIs (see Chap. 4).

The package of climate and energy policies was negotiated and adopted in parallel: the revised EU ETS, the Effort Sharing Decision for the non-ETS sectors, the Renewable Energy Directive and the CCS Directive for safe and secure storage of carbon which aimed at reducing regulatory risks for investors. Chris Davies, the Parliament rapporteur for the CCS Directive dossier, was highly supportive of accelerating CCS demonstration and commercialization. In cooperation with Avril Doyle (rapporteur for the proposal to revise the ETS), the two Members of Parliament (MEPs) presented a combined solution to use allowances from ETS to fund low-carbon technologies (Chiavari 2010; Åhman et al. 2018). After long and complex negotiations, a compromise was reached on a new mechanism that would set aside 300 million allowances from the trading system's New Entrance Reserve (NER 300) to co-finance up to 12 pilot CCS plants and new renewable-energy demonstration projects. This created a new EU funding programme independent of the FPs that could potentially fund the SET-Plan industrial initiatives. Member-states were also recommended to devote at least 50 per cent of the revenues from auctioning allowances to climate solutions, including demonstration projects under the framework of the SET-Plan.

By December 2008, the EU had put in place an unprecedented package of policies for stimulating demand and facilitating supply of low-carbon technologies. The market-pull part of the package was adopted as legally binding climate and energy policy instruments, but not the technology-push part represented by the SET-Plan. The 2008 Council conclusions and Parliament resolution were not transferred into a final legal text. Council and Parliament discussions revealed broad political backing for the SET-Plan, but also equivocal signals as to future direction, priorities and governance.

3.1.4   Conclusions

The Commission received wide political support for creating a low-carbon technology-push policy within the wider EU climate and energy market-pull package adopted in 2008. Also, generally welcomed were higher budgets for low-carbon energy research and innovation, and better coordination and prioritization of such efforts. However, the European Parliament and the member-states differed in their preferences concerning SET-Plan governance, the criteria for selection of priorities and specific technologies. Using the framework set out in Chap. 2, we can draw three main conclusions.

First, the Commission was not given any guarantee or legal means to ensure that the proposed technology priorities and greater focus on industrial demonstration projects would be supported by increased funding.

The European Parliament was not included as part of the Plan's governance system. Coordination of resources around priorities would be voluntary. EU energy R&I policy would in principle continue as a shared competence area for the EU and the member-states, but the Commission was not provided with formal authority to execute the EU's part of this shared competence. Second, the selection of technologies as EIIs meant a considerable divergence from the principle of technology neutrality.[25] The SET-Plan thus became vulnerable to mobilization and political conflict. Finally, the SET-Plan was not finalized in a binding text that could indicate a compromise between the equivocal signals from the Council and Parliament as to future direction, priorities and governance. This increased the risk that the adopted SET-Plan might get derailed due to further contestation in the implementation stage.

3.2   Implementing the Plan

Implementation of the SET-Plan commenced in 2008. All the new arenas for collaboration were established before 2009: The Steering Group, the European Industrial Initiatives, the European Energy Research Area and the new information and communication system (Research Council of Norway 2011). In spring 2008, the Steering Group for the SET-Plan was constituted by a first meeting between member-state and Commission representatives, to agree on joint programmes, EIIs and funding. A group of representatives from national ministries and funding agencies would, in liaison with the Commission, meet regularly to prepare Steering Group meetings. EII teams were constituted around the six technology priorities, and in 2009, these teams published more detailed master plans for realizing the EIIs in a series of Technology Roadmaps (Commission 2009a). The EERA launched the first joint programme in 2010. SETIS was constituted within the Commission's Joint Research Centre to update R&I capacities and technology maps regularly, keeping track of SET-Plan implementation.

The EII Technology Roadmaps specified goals and resource needs towards 2020 and beyond, envisioning how innovation and market deployment would follow from research and demonstration activities in

---

[25] The EU ETS was the only technology-neutral market-pull instrument in the climate and energy policy package.

the six technology areas. The roadmaps, drawn up by the Commission based on industry proposals, were advanced through discussions, workshops, multilateral meetings and expert consultations between the Commission services and the European energy technology platforms, the relevant sector associations, the research community, member-states and other stakeholders (Commission 2009f, p. 5). Great variation became evident among the six roadmaps as regards the targets formulated for how the EIIs were to contribute towards the 2020 climate and energy targets, in 'technology penetration' timelines, number of demonstration plants to be in place by 2020 and costs (Table 3.1). According to the Commission, the variation in targets reflected differing industry levels of ambition and vision (Commission 2009f, p. 6). More detailed implementation plans would be made before the final launch of the EIIs, first for the period 2010–2012 and then for 2013–2015.

3.2.1 Funding

The Commission's funding plan was outlined in the communication 'Investing in the Development of Low Carbon Technologies' (Commission 2009c). In 2007, before the SET-Plan was adopted, an estimated €2.38 billion (EU, national and industry funding) was allocated quite evenly to the six technology areas covered by EIIs.[26] On the basis of the Technology Roadmaps, the Commission estimated that €53 billion in fresh funding would be needed for the 2010–2020 period to realize the six EIIs. Moreover, €5 billion from EU and national budgets would be needed to realize the EERA joint programmes underpinning the EIIs (Commission 2009b).

The first main extension of the SET-Plan was proposed in this funding plan: a Smart Cities Initiative to demonstrate energy-efficiency technologies. Unlike the other EIIs that targeted major energy companies and utilities as partners for large-scale demonstration projects, the new initiative included 25–30 selected cities in Europe as partners, for demonstrating low-carbon energy transition for buildings, energy networks and

---

[26] A Commission staff working document underpinning the funding plan analysed the EII-technologies and those for which a dedicated European programme already existed. The Commission chose to call both groups of technologies, including fuel cells and hydrogen and nuclear fusion, 'SET-Plan priority technologies' (Commission 2009c. p. iii). In 2007, the 'fuel cells and hydrogen' JTI, not included among the EIIs, received €616 million in R&I support (Commission 2009b).

Table 3.1 Summary of Technology Roadmaps

EIIs	Main goals	Targets	Development needs	Demonstration projects	Costs to 2020
Wind	Offshore wind Grid integration	20% EU electricity consumption by wind	Tech for grid integration Large turbines offshore and onshore	10 large-scale (10–20 MW) prototypes by 2020	€6 billion
Solar PV and CSP	Penetration in urban areas/greenfield locations Grid integration Demonstrate competitiveness/readiness for mass deployment	12% EU electricity consumption 3% of EU electricity consumption	Sub-technologies and mass-production Up-scale to pre-commercial or commercial level	5 pilot mass-production manufacturing plants and a portfolio of demo projects 10 first-of-a-kind industrial-scale plants	€9 billion €7 billion
Electricity Grid	Enable additions of renewable energy	Grid integration of 35% renewables	Network flexibility and pan-European grid architectures.	20 large-scale projects	€2 bill. (excl. major costs for network operators)
Bioenergy	Widespread sustainable exploitation of biomass resources	14% bioenergy in EU energy mix, 60% GHG emission savings for biofuels	Along the whole value chain Advanced biofuels and biorefining	30 plants at a scale appropriate to level of maturity. Cellulostic biofuels demos expected online by 2010 and biorefineries by 2015)	€9 billion
CCS	Demonstrate commercial viability and deployment in coal-fired power plants by 2020		Up-scale tech to deal with costs and safety concerns	A portfolio of demonstration projects	€10.5 to €16.5 bill. (depending on number of demos)
Nuclear Fission	Generation-IV reactor to maximize safety and reduce radioactive waste	Sustain 30% share for nuclear power in the long run	Develop two new reactor concepts	Operation of two reactors by 2020	€5–€10 billion

Source: Based on Commission (2009a, f)

transport systems. It was estimated that €11 billion would be needed to realize this initiative (Commission 2009c).

To realize all the priorities of the SET-Plan and continue funding other energy R&I tasks, the Commission estimated that annual budgets of €8 billion per year would be needed (up from 3 billion in 2007). Most of this was expected to come from corporate actors (70 per cent share of total funding in 2007). However, the public authorities should increase their share from 30 per cent, as the economic crisis had made industry more cautious about investing in riskier technologies. The Commission also wanted an increase in the share of funds allocated at the EU level (20 per cent of total public research funding in 2007) or, alternatively, through joint programmes between member-states (Commission 2009c).

Specific funding options included national and EU programmes, some ongoing and some new. The Commission would utilize the new opportunities from the NER 300 Programme through which member-states would select projects based on criteria defined at the community level (Åhman et al. 2018). Other EU-level options included the EU Framework Programmes for research and innovation (FP) and the parallel EURATOM Programme to fund the nuclear power EII, and the European Energy Programme for Recovery (EEPR), adopted in response to the economic crisis.[27] Another important mechanism for funding would be public–private partnerships (combinations of grants, loans and loan guarantees to trigger industry investments), where the European Investment Bank (EIB) would have a pivotal role. The EIB had been granted extended rights to issue loans as part of the response to the economic crisis.[28] Several joint Commission–EIB financing facilities had already been developed, such as the Risk-Sharing Finance Facility co-funded under FP7 (Commission 2009c).

From October 2010 to March 2011, the funding plan was discussed, receiving the support of the Council and the European Parliament. However, the MEPs disagreed with the Plan's low-carbon technological priorities and its bias towards large-scale demonstration projects (see Chap. 4).

---

[27] The Intelligent Energy for Europe programme was envisaged to support the Smart Cities and Communities Initiative (SETIS Undated).

[28] Its Board of Governors (the finance ministers of the EU member-states) had decided to extend its budget for energy projects from €6.5 billion in 2008 to €9.5 billion in 2009 and €10.25 billion in 2010 (Commission 2009c).

3.2.1.1 The Framework Programme: FP7 (2007–2013)

The budget and main lines of allocation for FP7 had been settled before the SET-Plan was adopted. FP7 nevertheless was in line with the general objectives adopted for the SET-Plan, in underscoring the need for R&I to provide low-carbon solutions to address climate change, improve security of supply and increase the competitiveness of Europe's industries (European Parliament and the Council 2006). Thematic priorities determined for FP7 excluded oil and gas technologies, whereas a broad portfolio of low-carbon technologies had been included—far broader than the SET-Plan priorities (Advisory Group on Energy 2006). The Commission was to have some leeway to further align the FP closer to SET-Plan priorities, through its central role in defining specific work programmes and calls for projects based on advice from member-states and stakeholders (Andrée 2009). However, ex-post assessments indicate that FP7 was finalized as initially planned, supporting a far broader portfolio of low-carbon energy technologies than envisaged under the SET-Plan (Ricardo Energy & Environment 2017; van der Veen et al. 2014), (see Table 3.2).

FP7 contributed to raising EU energy R&I budgets as compared to FP6, including for the SET-Plan technologies. Compared to the level of resources required to realize the EIIs, however, FP7 funds would only cover a fraction of the documented needs (see Table 3.1).[29]

Table 3.2 FP7 budget breakdown

Category	Thematic area	Projects (N)	% Split of budget
Renewable energy	Bioenergy	60	42% renewables
	Wind	27	
	Heating and cooling	25	
	Ocean	12	
	Hydropower	2	
	Concentrated solar power	18	
	Photovoltaic power	30	
	Future emerging tech/materials	29	
Fuel cells & hydrogen		8	15%
Smart grids		58	15%
CCS/clean coal		59	10%
Energy efficiency		40	13%

Source: Based on Ricardo Energy & Environment (2017, Table 15)

[29] Annual energy budgets up from around €140 mill/year in FP6 to ca €335 mill/year in FP7 (van der Veen et al. 2014. p. vi).

The SET-Plan EIIs envisaged major resources for large-scale industrial demonstration projects, whereas the share of FP7's energy budget for this purpose was only marginally up-adjusted as compared to FP6 (van der Veen et al. 2014). A specific mechanism adopted under FP7 to spur European Investment Bank co-funding of industrial demonstration projects, the Risk-Sharing Finance Facility, had yielded meagre outcomes for energy projects specifically.[30] Energy demonstration projects realized under FP7 were mostly small-scale, in the range of up to €15–20 million, with only a few large-scale exceptions for the EII technology areas, Smart Grids, Bioenergy and Wind Power (van der Veen et al. 2014, p. 23).[31]

The SET-Plan further envisaged upscaled member-state coordination of national funding programmes around the selected priorities. FP7 aimed at contributing by continuing the ERA-NET mechanism to incentivize member-state joint actions. However, in the area of energy technology, this mechanism attracted scant coordination on the part of the member-states: only a few participated in joint actions, and the level of coordinated funding increased only slightly from FP6 to FP7 (Commission 2014a).[32]

The Knowledge and Innovation Community Programmes initiated by the European Institute of Technology and aimed at establishing joint educational programmes, coordination of research and industrial innovation projects, were another FP7 mechanism that could deliver SET-Plan implementation. A separate KIC on energy, *InnoEnergy*, was established in 2010—however, with a considerably broader thematic focus than the initial SET-Plan (InnoEnergy 2018).

Thus, FP7 made only limited contributions to funding the SET-Plan and its priorities, specifically co-funding large-scale demonstration projects close to market deployment. Industrial demonstration projects funded under FP7 were generally small-scale, with a few large-scale exceptions aligned with the SET-Plan.

---

[30] This is indicated by the reduction for the share of funds allocated to energy projects. While energy projects had a share of 48 per cent of projects awarded funds in 2007, the share had dropped to 14 per cent by 2012 (European Investment Bank 2007, 2013. p. 8).

[31] The costliest projects had budgets of up to €200 million (van der Veen et al. 2014, p. 23).

[32] The average funding per year was €16.6 million under FP6 and €18 million under FP7 to joint energy-related activities (Commission 2014a).

3.2.1.2   FP8: Horizon 2020 (2014–2020)

Between 2011 and 2013, the follow-up FP-programme, FP8 or Horizon 2020, was developed, providing new opportunities for better aligning the FP with the SET-Plan. Kick-starting the process of establishing FP8, a 2011 Green Paper of the Commission emphasized that the low level of private funding remained a bottleneck for research and innovation in Europe. More resources under FP8 would need to be combined with EIB funds to leverage private industry co-funding of R&I (Commission 2011a, p. 10). Such a reorientation would be in line with the SET-Plan aim of spurring large-scale industrial demonstration projects.

However, stakeholder responses to the Green Paper revealed disagreement on FP8 reorientation (Commission 2011b). Illustrative is the position paper submitted by the SET-Plan's EERA, which recommended channelling more FP8 resources to basic and applied R&D and not to industrial demonstration projects—allegedly because the scope for cost reductions would be greatest for projects in the early stages of development (EERA, October 2010). This document indicated tensions inherent in the SET-Plan on the relative distribution of funds for EERA joint basic research at universities and support to industrial demonstration of more mature technologies within the EIIs.

The European Parliament's resolution on the Green Paper ambiguously supported more resources to industrial research and innovation. The Parliament feared that too great a focus on large-scale demonstrations in the FP8 would favour large companies, at the expense of SMEs and research institutions interested in a range of less-mature technologies (European Parliament 2011).

The Commission's proposal for FP8 aimed at balancing these concerns by distributing funding to achieve excellence in the science base; to tackle societal challenges; and to create industrial leadership and boost competitiveness. The proposal included energy projects in all three budget posts (Commission 2011c), and the SET-Plan was specifically mentioned. It was proposed to fund demonstration projects by an upscaled Risk-Sharing Finance Facility based on boosting the EIB and industrial co-funding. 'Secure, clean and efficient energy' was outlined as one of six societal challenges and budget post for supporting the implementation of the SET-Plan and EIIs. This said, the proposed list of technology priorities was only partly aligned with the SET-Plan—far more low-carbon technology areas and solutions were listed as eligible for support.

The European Parliament co-decided with the Council on the broad content and scale of Horizon 2020. The total budget adopted (€77 billion) was in line with the Commission proposal. The Council had opted for a 10 per cent cut and the European Parliament a 30 per cent increase (European Parliament 2013a). The European Parliament wanted more of the FP8 budget to be allocated to energy research than did the Council, with a middle-ground settlement reached in the trilogue negotiations (Council 2013a). As such, the European Parliament fronted strengthening energy R&I budgets at the EU level, one of the objectives of the SET-Plan.[33]

A full ex-post assessment of alignment between Horizon 2020 and the SET-Plan will not be possible until the programme ends in 2020. However, several indicators of non-alignment can be observed already. First, the work programmes adopted and planned under Horizon 2020 have had a considerably broader focus than the 2008 SET-Plan in terms of technology areas targeted for support.[34] The work programmes reflected the list of priorities proposed for FP8 by the Commission and decided by the Council and Parliament in 2013. This indicates that there had been no comprehensive efforts at aligning Horizon 2020 more closely with the initial SET-Plan priorities in the 'work programming' stage.

Secondly, specific funding mechanisms in Horizon 2020 that could contribute to alignment with the SET-Plan were not fully applied for this purpose. One mechanism, continued from FP7, was the Risk-Sharing Finance Facility, renamed InnovFin Energy Demo Projects (EDP), to target the funding of industrial demonstration projects (SETIS 2015). However, this mechanism was adopted with eligibility criteria that did not match with the initial SET-Plan priorities: broader areas of renewable energy technology were made eligible for support, as were fuel cells and hydrogen technologies. The first project awarded funds under EDP in 2016 was in ocean energy, not an initial priority of the SET-Plan (European Investment Bank 2016).[35]

---

[33] The compromise entailed that around €5.9 billion would be allocated to energy research and innovation (Council 2013a).

[34] See Commission Horizon 2020 info pages, Secure, Clean and Efficient Energy, http://ec.europa.eu/programmes/horizon2020/en/h2020-section/secure-clean-and-efficient-energy, accessed 14 June 2018.

[35] In 2017, a solar power and a bioenergy-based heat- and power-demonstration plant were funded under the mechanism (Commission 2017c).

Another mechanism, also continued from FP7, the ERA-NET mechanism, aimed specifically at spurring member-states to coordinate their national funding programmes. The number of ERA-NET joint actions on energy between the member-states did indeed increase as compared to FP7, encompassing all SET-Plan technology fields and specifically targeting large-scale demonstration projects (*SETIS Magazine*, April 2017a). However, ERA-NET joint actions also included a wider scope of technologies than those initially selected for the SET-Plan, including on geothermal and ocean energy (*SETIS Magazine*, November 2017b). Experience from calls on joint action indicate meagre interest from industry and lack of investment decisions for demonstration projects that had been awarded funds. From 2017 onwards, the scope of energy ERA-NET topics has been widened beyond demonstration projects.

The result was rather weak alignment between priorities in the *initial* SET-Plan and funding under Horizon 2020 for the period 2014–2020. Political compromises meant that available funds, as for the former FP7, were allocated to a far broader portfolio of technologies than envisaged by the initial SET-Plan EIIs. This contrasts strongly with Commission statements on its Horizon 2020 info pages for work programmes in the field of energy: 'Since 2008, the SET Plan has been the centre-piece of our research and innovation policy in the field of energy. It is the reference point for European, national, regional and private investment.'[36] However, this statement makes sense in light of the 2015 reforms of the SET-Plan. An important part of the reform involved adding new technology areas as priorities to the SET-Plan, so the post-reform Plan was considerably more in line with priorities already decided for Horizon 2020 (see below).

3.2.1.3 EURATOM

The EURATOM Programme was to fund nuclear power R&I, including the specific SET-Plan EII on Sustainable Nuclear Power. Like the broader FP7, however, the EURATOM research programme for 2007–2011 was adopted before the SET-Plan and was thus not aligned with it. The programme targeted research on a far broader portfolio of priorities than

---

[36] Commission Horizon 2020 info pages, Secure, Clean and Efficient Energy, http://ec.europa.eu/programmes/horizon2020/en/h2020-section/secure-clean-and-efficient-energy, accessed 14 June 2018.

what had been selected as key priority for the nuclear EII, which was accelerated development and demonstration of new Generation IV fission reactors (see Table 3.1 above). The bulk of the €2.7 billion EURATOM budget (70 per cent) was allocated to fusion research.[37] Fission research was by comparison allocated around €100 million a year, with only part of this sub-budget devoted to new Gen IV reactor designs (Commission 2017d). This meant poor alignment between the SET-Plan Sustainable Nuclear Power EII and EURATOM funding for the period 2007–2011. The Commission improved alignment with the proposed EURATOM extension programme 2012–2013 by including demonstration of Gen IV fission reactors in the proposal (Commission 2011d, p. 7).

Political discussions in the Council reflected conflicting views among member-states on whether the programme should focus on safety improvements of existing plants, or also the development of a new generation of fission reactors as envisaged by the SET-Plan EII (Barnes and Barnes 2018). Discussions were affected by the concurrent (March 2011) accident at the Japanese Fukushima plant. In the end, the Council arrived at a compromise that entailed reorienting EURATOM towards exclusively funding the safety aspects of nuclear power (Austrian Ministry of Science and Research 2012). The bulk of the budget (86 per cent) was allocated to nuclear fusion. The minor share allocated for nuclear fission projects would be allocated to existing nuclear-plant safety projects and the development of new Gen IV reactors (Council 2011). As funds available from EURATOM was restricted, realizing the SET-Plan EII would depend heavily on member-state and industrial funding.

The Council's decision on the next EURATOM extension, for 2014–2018, entailed a reduction in budget compared to the previous five-year period. The focus for fission research in the programme was on environmental safety and security regarding *current* nuclear technology (Commission 2017e, p. 19). This focus again limited the opportunities for realizing the SET-Plan Sustainable Nuclear Power EII, and the initial plan for realizing Gen IV fission reactor demonstration projects by 2020, outlined in the Technology Roadmap, was postponed.[38]

---

[37] Included was support to the major international ITER fusion power demonstration project under construction in France.

[38] 2025 was set as new deadline for the pilot and demonstration plants in the Sustainable Nuclear Power EII Implementation Plan for 2013–2015 (SNETP Undated, p. 3).

3.2.1.4 European Energy Programme for Recovery

The EEPR Regulation was adopted by the Council and Parliament in June 2009 (European Parliament and the Council 2009). The initial Commission proposal made no mention of the SET-Plan: support was envisaged for investments in energy infrastructure, energy efficiency, clean construction and automobile technologies (Commission 2008b). The European Council requested better alignment with the SET-Plan and for funding demonstration of CCS and offshore wind projects in addition to energy infrastructure; this was followed up by the Commission in its proposal of January 2009 (Commission 2009d). The bulk (60 per cent) of the €3.98 billion budget would be allocated to projects applying already well-known gas and electricity infrastructure technologies, thus not related to the SET-Plan. Forty per cent would be allocated to SET-Plan demonstration projects: offshore wind (14 per cent) and CCS projects (26 per cent) (Commission 2009e). The European Parliament opted for reinserting local-level energy efficiency and renewable energy projects as eligible for support, initially envisaged by the Commission (European Parliament 2010, p. 15). A compromise text was added after trilogue negotiations: unused funds for wind and CCS could be allocated to efficiency and broader renewable energy after scheduled assessments in 2010 (European Parliament 2010, p. 15).[39]

Eligible projects had already been listed in an annex to the EEPR Regulation: 5 offshore windpower projects shared by 10 member-states and 13 CCS projects in 7 member-states. The award decision of 9 December 2009 gave support to nine offshore windpower projects and six CCS projects.[40] Available resources were thus limited: around €1 billion, to be divided among six large-scale CCS projects and €550 million for nine offshore windpower projects (Commission 2009e).

Between 2013 and 2018, EEPR implementation showed mixed results. The programme had mainly contributed to realization of gas and electricity energy infrastructure projects that were not part of the SET-Plan (Commission 2018a). Many SET-Plan relevant projects had been cancelled before the final investment decision. The greatest problems were

---

[39]€115 million in non-used funds was made available to a wide range of energy efficiency and renewable energy solutions at local level (European Parliament and the Council 2010).

[40]Windpower projects in Belgium, Denmark, Germany, Ireland, the Netherlands, Norway, Sweden and the United Kingdom; CCS projects in Germany, Italy, the Netherlands, Poland, Spain and the United Kingdom (Commission 2009e).

recorded for CCS projects. Of the six awarded projects, no demonstration plants had reached final investment decision, and only one smaller project had been completed (Commission 2018a). The situation for offshore windpower projects was better: four out of nine awarded projects had been completed, two had been terminated and three were ongoing (Commission 2018a).[41]

Thus, the EEPR became partly aligned to the SET-Plan by the European Council. However, the fund failed to attract industry investment in large-scale CCS demonstration projects relevant for the SET-Plan. The offshore wind demonstration plants co-funded by the programme fared better, although some projects were cancelled, and others postponed (Commission 2018a). However, SET-Plan non-related gas and electricity infrastructure projects did well.

3.2.1.5 NER 300

The NER 300 Programme, focused on CCS and renewables, represented another possibility for realizing the SET-Plan EIIs. According to the eligibility criteria adopted in 2010, each member-state should be entitled to propose at least one project in line with its own priorities and national projects in the pipeline. Renewable energy projects should not only be large scale (Åhman et al. 2018).

The programme, developed by DG Environment/DG Climate Action, was not systematically coordinated with the SET-Plan as indicated by the broader portfolio of technologies eligible for funding. Most projects applied for were in line with the SET-Plan EIIs but included other, smaller, projects and technologies as well, such as ocean and geothermal energy (Åhman et al. 2018).

NER 300 quickly ran into serious trouble: a plunge in EU ETS allowance prices reduced the fund from expected €9 billion to around 2.1 billion, to be awarded to 42 projects in two calls and co-funded by national programmes. Awards were continuously delayed, and many projects never reached a final investment decision. Of 23 projects awarded funding under the first call, only three were operational by June 2017; seven projects had been withdrawn (Åhman et al. 2018).

---

[41] Projects have focused on large-scale testing, manufacturing and deployment of innovative turbines and offshore foundation structures (six projects), and development of module-based solutions for the grid integration of large amounts of wind electricity transmission (three projects) (Commission 2018a).

The eligibility criteria entailed that only a limited number of large-scale projects could be realized without the member-states and industry massively increasing their commitment to funding (Åhman et al. 2018). An additional problem came from the limitations as to how NER 300 funds could be combined with other EU-level programmes. Co-funding with the EEPR was allowed—but funding from the NER 300 would be reduced by any amount granted by the EEPR, which constrained combining EU-level funds to create a larger critical mass of resources (Commission 2010).[42] CCS projects were very large in scale, and failed completely: none of the 13 projects under the first call were awarded funding. Large-scale biofuel projects also encountered problems, and five projects were withdrawn from the first call. Projects with successful awards and final investment decision included wind projects, small- to medium-scale bioenergy projects, as well as ocean and geothermal projects—the latter two areas not among the initial priorities for the SET-Plan EIIs (Åhman et al. 2018).

3.2.2   Reform

In 2013, the Commission initiated a SET-Plan reform process (Commission 2013a, p. 2). The reform was motivated by implementation problems experienced, emerging new climate and energy policies for 2030 and the need for new solutions to energy-system challenges. These challenges emerged from the massive deployment of decentralized and variable supply of solar and windpower.

The Commission proposed a wide extension of priorities. Innovative solutions for the entire energy system would be added; the EII initiatives on specific energy supply technology areas would continue, extended by the inclusion of ocean energy, heating/cooling, and energy storage—reflecting actual R&I priorities and funding (Commission 2013a, p. 8). This broadening of priorities signalled a fading of the Commission's original ambition to pick winners (*Energy Post*, 31 July 2013). Opening R&I programmes to more technologies was justified with reference to the generally long lead-times and uncertainties as to which technologies would finally succeed in global energy markets beyond 2020; it was proposed that large-scale projects beyond what member-states could achieve alone should be kept as a key selection criterion for the SET-Plan (Commission 2013a).

---

[42] Article 2 (3) in the Commission Decision laying down criteria and principles for NER 300 (Commission 2010).

Also proposed was strengthening SET-Plan governance: this reflected the implementation problems thus far. The Steering Group would need a new mandate that gave it more authority. The EIIs should adjust their structure and increase industry participation.[43] Also new stakeholders should be brought into the EIIs: local actors, SMEs, regulators, network operators, financiers and consumer organizations. The EERA should step up its collaboration with the EIIs. Strengthened R&I capacities should (again) serve to increase global market opportunities for European industries. What was not clear was how these changes were to be realized, and what consequences they would have for who should govern the implementation of the Plan.

To kick-start the reform, the Commission proposed an *Integrated Roadmap* as a master plan for realizing new priorities. The roadmap should cover the entire research and innovation chain from basic research to demonstration and identify clearer roles and tasks for stakeholders. It should be developed under the guidance of the SET-Plan Steering Group by the end of 2013. The member-states and the Commission should jointly develop a new and more concrete action plan by mid-2014.

In June 2013, the Council broadly endorsed the reform (Council 2013b). A European Parliament Committee report submitted prior to the Commission proposal welcomed the reform and called for a stronger focus on energy efficiency (European Parliament 2013b, p. 15). Successive online consultations attracted responses from 150 stakeholders and revealed highly diverse opinions on priority areas for a reformed SET-Plan (Interview B).

The reform was considerably delayed. In December 2014, the Joint Research Centre published a more detailed document outlining main directions, anchored in the SET-Plan Steering Group and the October 2014 European Council's decision on new 2030 climate and energy targets (Commission 2014b).[44] Changes in governance were signalled, to improve implementation—subgroups to the SET-Plan Steering Group should be established, with more limited representation among the most committed member-states (*SETIS Magazine*, November 2017b).

---

[43] The EIIs would later merge with the Energy Technology Platforms to create new ETIPs (European Technology and Innovation Platforms).

[44] At least 40 per cent reduction in GHG, 27 per cent share for renewable energy (binding only at EU level) and an indicative energy efficiency target of 27 per cent that would be reviewed in 2020 with a view to an increase to 30 per cent.

Indeed, commitment to the Steering Group had been a major problem (see Chap. 4).

The reform indicated a reorientation and massive widening of R&I efforts compared to the initial SET-Plan. New priorities would include solutions for strengthening consumer engagement in the energy transition, electrification of new sectors, application of new energy-storage solutions and the development and commercialization of low-carbon energy supply technologies. Thirteen specific R&I themes were singled out, reflecting the new priorities but also covering the SET-Plan's existing technology EIIs as well as new ones in the technology areas of ocean energy, geothermal energy and heating/cooling. These themes would constitute the basis for the action plan that the member-states should apply as guidance for future coordination of national R&I programmes.

In 2015, the new Juncker Commission proposed new SET-Plan priorities in its Energy Union strategy (Commission 2015a).[45] The new R&I strategy envisaged four main priorities, later presented as 'a more targeted approach' (Commission 2015b, p. 8):

- renewable energy technologies, including environment-friendly production and use of biomass/biofuels and energy storage
- smart grids, smart home appliances, smart cities, and home automation systems
- efficient energy systems and technology to make the building stock energy-neutral
- more sustainable transport systems.

Two initially prioritized SET-Plan technology areas were given lower priority: CCS and nuclear power technologies were given status only as *additional priorities* that could be promoted by collaboration between the Commission and interested member-states. CCS and CCU (carbon capture and use) for the power and industrial sectors were still deemed critical for achievement of the 2050 climate objectives, as was continued technological leadership in nuclear power, which was important to avoid a further increase in energy and technology dependence. The proposal reflected the major challenges experienced in connection with realizing the EIIs on

---

[45] The Energy Union proposed reform along five climate and energy policy dimensions, one covering research, innovation and competitiveness.

CCS and sustainable nuclear power, due to the lack of commitment of the member-states and industry to joint funding of large-scale industrial demonstration plants.

In September 2015, the Commission formally presented the reformed SET-Plan based on the 4+2 priorities, anchored in the SET-Plan Steering Group and inputs from the European Council (Commission 2015b). Ten 'actions' were specified for the six priority areas and potential funding sources linked broadly to the SET-Plan and other Energy Union innovation priorities. In 2017, the Commission presented a different framing of the SET-Plan—several priorities were grouped as covering 14 'energy sectors' for which implementation plans would be developed (Commission 2017a, p. 19). This recurrent reframing of themes under broad headings and changing the number of actions reflected the EU's problems in agreeing on priorities.

The Commission started to implement the reform by drafting a series of 'issue papers' on the various actions. Based on Declarations of Intent between research and industry stakeholders and interested countries, the 11 first implementation plans had been adopted by June 2018 (SETIS 2018). With the reform process, the SET-Plan drifted further and further away from its original idea of concentrating resources for only a few specific technology areas and project types (demonstration projects) as preferred procedure for accelerating the market uptake of new low-carbon innovation.

3.2.3   Concluding Assessment

The SET-Plan had been adopted with the objectives of strengthening, focusing and giving coherence to low-carbon energy R&I efforts in Europe, thereby accelerating innovation towards 2020 and beyond with large-scale demonstration projects for selected technologies as a key instrument. To achieve these objectives, the Plan assigned various tasks to actors, with shared responsibilities for implementation. These tasks included: consolidating selected technologies and providing detailed implementation plans for the priorities; providing and coordinating the resources required to realize these priorities by ensuring that funds (at EU, national, and industry levels) were aligned with the Plan; and, realizing the demonstration projects stipulated by the Plan. These tasks and objectives have been accomplished only to a limited extent.

The first task was to be executed by EII teams and the EERA within the six initial technology priorities. The EII teams progressed by adopting Technology Roadmaps for the six EIIs in 2009, preparing their official launch in 2010. Next, more detailed two-year implementation plans were delivered for the periods 2010–2012 and 2013–2015, and the specific Smart Cities and Communities Initiative was launched in 2011. Assessments indicate mixed results. On the one hand, the EII Teams, supported by European Technology Platforms, 'demonstrated capacity to prioritizing and planning of actions'; however, the teams were narrow in scope, lacking a 'balanced and representative group of industries and often of Member States with clear commitment to strategic planning', curtailing the process of identifying and prioritizing actions (Commission 2013b, p. 8).

European research institutions organized in EERA made progress by defining and planning joint research programmes. The final selection of such joint programmes was taken by an EERA Executive Committee among European leading energy research institutions, supported by an EERA Secretariat. The first seven programmes, launched in 2010, focused on geothermal energy, wind energy, photovoltaics, smart grids, nuclear-power materials, bioenergy and CCS. In 2011, six more joint programmes were launched.[46] From 2013, several others were added.[47]

The EERA joint programmes adopted show a clear mismatch with the SET-Plan, first and foremost because of the far broader portfolio of technologies targeted for joint research efforts. Moreover, the programmes, even in technology areas selected for the SET-Plan, were poorly aligned with the implementation plans of the EIIs (Commission 2013b, p. 6). In the end, the SET-Plan expanded to include technologies and solutions for 14 energy sectors by 2017.

The second main task involved providing the funds needed to realize the SET-Plan, estimated by the Commission in 2009 at €8 billion/year (Commission 2009c). Energy R&I data collected by SETIS indicate an increase in total annual EU funding for SET-Plan technology areas between 2007 and 2011.[48] Comparable data for later periods have not

---

[46]'Advanced Materials and Processes for Energy Applications'; 'Concentrated Solar Power'; 'Energy Storage'; 'Fuel Cells and Hydrogen'; 'Ocean Energy' and 'Smart Cities'.

[47]'Environmental, Economic and Social Impact Analysis' and 'Shale Gas' in 2013, 'Energy Efficiency in Industry' (2015), 'Energy Systems Integration' (2016). It was later decided to stop the 'Shale Gas' joint programme.

[48]Estimated at €3 billion allocated in 2007, €5 billion in 2010 and €7 billion/year in 2011 (JRC 2015).

been published by SETIS, but a new estimate for the period 2010–2015 indicates an 8 per cent increase in the total EU energy R&I budget for the larger number of priorities in the revised SET-Plan and broader Energy Union priorities (Commission 2017a). SETIS data certainly indicate an increase in total EU funding of energy R&I after the launch of the SET-Plan: however, they only provide estimates and not an exact picture of how total funds were allocated to individual EII priorities (JRC 2015; Commission 2017a). This reflects lack of reporting on energy R&I data by the member-states and industry.

Despite uncertainties, SETIS-collected data indicate that industry funding remained the key source for all the SET-Plan technology areas except nuclear power and CCS technologies (Commission 2017b). Data further indicate relatively high levels of funding by industry in technology areas initially excluded from the SET-Plan but added after the 2015 revision, like energy efficiency in industry and smart solutions for consumers (Commission 2017a). This indicates considerable non-alignment with the initial SET-Plan priorities, as well as later adaptation of priorities in the SET-Plan to actual on-the-ground funding patterns.

Because of the financial crisis, the Commission in 2009 expected a falling share of total funding from industry, to be compensated by public funding at national and EU levels. And indeed, the industry share of total SET-Plan funding declined in the period 2007–2011 (JRC 2015, p. 4). Also, EU-level programmes saw a declining share, whereas member-state public programmes increased in relative importance. National programmes remained fragmented, however, as indicated by the scarcity of ERA-NET joint actions on energy under FP7 and the continued meagre outcomes of such joint actions under Horizon 2020.

Fragmentation and weak alignment with the SET-Plan remained a problem also for EU-level programmes. Initial SET-Plan priorities were reflected in the Framework Programmes only to a limited extent. A notable example from energy R&I programmes is the NER 300, which developed its own procedure for allocating funding even though it became operational after the SET-Plan. The EEPR, initially foreseen as independent of the SET-Plan, became linked after intervention by the European Council. EEPR and NER 300 saw some project coordination (three projects co-funded); however, additionality in funding was not allowed, which constrained opportunities for pooling of resources.

A final main task for implementing the SET-Plan was the actual realization of joint large-scale demonstration projects under EIIs—the core priority of the SET-Plan. To our knowledge, no full list of SET-Plan demonstration projects realized from the different EIIs has been recorded by SETIS. Our study of various relevant EU-level funding programmes and mechanisms, NER 300, the EEPR, EURATOM, the Risk-Sharing Finance Facility and ERA-NET joint actions under the FP, indicates that many large-scale demonstration projects failed to attract enough funding to reach final investment decisions by industry. A major Commission-initiated study from 2016 supports this interpretation: many large-scale low-carbon energy demonstration projects covering EII technology areas faced tremendous challenges in raising sufficient funding to achieve financial close, complete construction, become fully operational, and thereby prove operational performance for the market (Commission 2016).[49] National public co-funding remained weak—only a few national funding programmes in 2015 focused specifically on supporting mature-technology demonstration projects, whereas most programmes would not fund projects of high maturity (Commission 2016, pp. 12–14).[50]

We have also observed considerable variation among the SET-Plan technology areas, indicating that some were poorer served than others in the funding landscape (Commission 2016). The record has been very poor as regards the realization of CCS pilot and large-scale demonstration projects and Sustainable Nuclear Power EIIs. No pilots or large-scale CCS demonstration projects have been realized, and the construction of large-scale demonstration projects for Generation IV nuclear reactors has been delayed. The best results have appeared in the technology areas of bioenergy, solar and windpower (Commission 2016, p. 99). However, even within these areas, the results are mixed if compared to the EII Technology Roadmaps and implementation plans. Funding mechanisms have apparently worked reasonably well in generating investments for the most mature solar PV power, windpower and mass-incinerated biomass heat

---

[49] Total resources available for such demonstration projects were estimated at around €4 billion, when measured across both EU support schemes (such as the NER 300 and InnovFin EDP) and member-state support schemes. This leaves a public funding shortfall of around €10 billion to achieve the maximum levels of demonstration projects, corresponding with the extended list of technologies under the revised SET-Plan (Commission 2016, p. 22).

[50] National funding programmes focusing specifically on full-scale demonstration projects at TRLs 7–8 were identified only in Denmark, Germany and the United Kingdom. France recorded the highest total level of funding for such projects among the member-states by a combination of programmes and loans. Norway and Sweden had programmes that provided some support (Commission 2016).

and power technologies (Commission 2016, p. 99). Achieving other planned EII outputs has been harder, as with large-scale second-generation biofuels demonstration plants, concentrated solar and offshore windpower demonstration projects (p. 145).[51] Mixed results have also been recorded for demonstration projects planned under the European Electricity Grid EII. There are many demonstration projects carried out at national level, whereas joint projects at the EU level have proven more difficult to fund and realize (Commission 2016, p. 145).[52] The Smart Cities and Communities Initiative had a late start; a total of 12 ongoing projects have been funded under Horizon 2020 since 2014 from three different calls, some defined as lighthouse projects, defined as large-scale demonstration projects[53] (Commission 2018b).

Summing up, then, the SET-Plan aimed at strengthening, focusing and giving coherence to EU energy R&I by selecting a portfolio of priority technology areas for cooperation and coordination of resources, involving demonstration projects within these areas. We have noted evident implementation problems and failures in the Plan as regards delivering on its promises. The level of energy R&I resources has increased, but not been well aligned with the initially selected priorities of the Plan. EU, national and industry funding programmes have not been adequately focused on the initially selected technology areas and large-scale demonstration projects (Commission 2016, 2017a). In 2015, the SET-Plan was revised to reflect the technologies that were funded: but this implied a loss of focus and continued fragmentation of resources rather than concentration and coherence. The revision extended the number of key priorities from the initial six technology areas to encompass 14 'energy sectors', masking a wide range of individual technologies. In the end, it was industry, national and EU R&I funding programmes and priorities that steered the SET-Plan, rather than the other way around.

---

[51] According to the 2017 implementation plan for the revised SET-Plan CSP initiative, 'almost none of the six CSP projects which were included in the final list of awardees under NER 300 has reached the financial close yet—and there are indications that some of them will not go ahead, meaning approximately 300 million EUR is currently frozen without the possibility to impact R&I in Europe for this sector (Initiative for Global Leadership in Concentrated Solar Power, November 2017, p. 11).

[52] This EII was coordinated by ENTSO-E, the pan-European TSO-organization, and became adapted to the R&I plan for this organization.

[53] €500 million in EU-level funding was reported for these projects (Commission 2018b).

REFERENCES

Advisory Group on Energy. (2006, November 8). *Opinion of the 2007 Workprogrammes.*   https://ec.europa.eu/research/fp7/pdf/old-advisory-groups/energy-wp-2007.pdf. Accessed 02 Oct 2018.

Åhman, M., Skjærseth, J. B., & Eikeland, P. O. (2018). Demonstrating Climate Mitigation Technologies: An Early Assessment of the NER 300 Programme. *Energy Policy, 117,* 100–107.

Andrée, D. (2009). *Priority-Setting in the European Research Framework Programmes.* VINNOVA, Swedish Governmental Agency for Innovation Systems.

Armstrong, K. A., Begg, I., & Zeitlin, J. (2008). 'EU Governance After Lisbon,' JCMS Symposium. *Journal of Common Market Studies, 46*(2), 413–450.

Austrian Ministry of Science and Research. (2012, February 12). *Austrian Position Paper on 'HORIZON 2020'.* https://era.gv.at/object/document/1601. Accessed 02 Oct 2018.

Barnes, P. M., & Barnes, I. (2018). *The Politics of Nuclear Power in the European Union. Framing the Discourse: Actors, Positions and Dynamics.* Opladen: Barbara Budrich Publishers.

CEPS. (2011). *The SET-Plan from Concept to Successful Implementation* (CEPS Task Force Report). Brussels: Centre for European Policy Studies.

Chiavari, J. (2010). The Legal Framework for Carbon and Capture and Storage in the EU (Directive 2009/31/EC). In S. Oberthür, C. R. Kelly, & M. Pallemaerts (Eds.), *The New Climate Policies of the European Union* (pp. 151–178). Brussels: VUB Press.

Commission. (1988). *The Greenhouse Effect and the Community: Commission Work Programme Concerning the Evaluation of Policy Options to Deal with the Greenhouse Effect.* COM (88) 656 Final, 16 November. Brussels: European Commission.

Commission. (2000a). *Towards a European Strategy for the Security of Energy Supply.* COM (2000) 769 Final, 29 November. Brussels: European Commission.

Commission. (2000b). *Towards a European Research Area.* COM (2000) 6, 18 January. Brussels: European Commission.

Commission. (2000c). *Making a Reality of the European Research Area: Guidelines for EU Research Activities (2002–2006).* COM (2000) 612 Final, 4 October. Brussels: European Commission.

Commission. (2003). *Investing in Research: An Action Plan for Europe.* COM (2003) 226 Final/2, 04 June. Brussels.

Commission. (2004). *Technology Platforms: From Definition to Implementation of a Common Research Agenda.* Brussels: European Commission Directorate-General for Research.

Commission. (2006a). *Green Paper on a European Strategy for Sustainable, Competitive and Secure Energy.* COM (2006) 105 Final, 8 March. Brussels: European Commission.

Commission. (2006b). *Summary Report on the Analysis of the Debate on the Green Paper 'A European Strategy for Sustainable, Competitive and Secure Energy'.* SEC (2006) 1500, 16 November. Brussels: European Commission.

Commission. (2007a). Limiting *Global Climate Change to 2 Degrees Celsius: The Way Ahead for 2020 and Beyond.* COM (2007) 2 Final, 10 January. Brussels: European Commission.

Commission. (2007b). *An Energy Policy for Europe.* COM (2007) 1 Final, 10 January. Brussels: European Commission.

Commission. (2007c). *Towards a European Strategic Energy Technology Plan.* COM (2006) 847 Final, 10 January. Brussels: European Commission.

Commission. (2007d). *SET-Plan Impact Assessment.* SEC (2007)1508/2. Brussels: European Commission.

Commission. (2007e). *A European Strategic Energy Technology Plan – Towards a Low Carbon Future.* COM (2007) 723 Final, 22 October. Brussels: European Commission.

Commission. (2008a). *Second Strategic Energy Review – An EU Energy Security and Solidarity Action Plan.* COM (2008) 744/3. Brussels: European Commission.

Commission. (2008b). *A European Economic Recovery Plan.* COM (2008) 800 Final, 26 November. Brussels: European Commission.

Commission. (2009a). *SETIS Summary of Roadmaps.* https://setis.ec.europa.eu/summary-of-roadmaps. Accessed 13 June 2018.

Commission. (2009b). *R&D Investment in the Priority Technologies of the European Strategic Energy Technology Plan.* SEC (2009) 1296, 07 October. Brussels: European Commission.

Commission. (2009c). *Investing in the Development of Low Carbon Technologies (SET-Plan).* COM (2009) 519 Final, 07 October. Brussels: European Commission.

Commission. (2009d). *Proposal for a Regulation of the European Parliament and of the Council Establishing a Programme to Aid Economic Recovery by Granting Community Financial Assistance to Projects in the Field of Energy.* COM (2009) 35 Final 2009/0010 (COD), 28 January. Brussels: European Commission.

Commission. (2009e). *Selection of Offshore Wind and Carbon Capture and Storage Projects for the European Energy Programme for Recovery.* MEMO/09/543, 9 December. Brussels: European Commission.

Commission. (2009f). *A Technology Roadmap.* Commission Staff Working Paper, SEC (2009) 1295, 07 October. Brussels: European Commission.

Commission. (2010). Decision of 3 November 2010 Laying Down Criteria and Measures for the Financing of Commercial Demonstration Projects that Aim at the Environmentally Safe Capture and Geological Storage of CO2 as Well as Demonstration Projects of Innovative Renewable Energy Technologies Under the Scheme for Greenhouse Gas Emission Allowance Trading Within the Community Established by Directive 2003/87/EC of the European Parliament and of the Council (Notified Under Document C(2010) 7499), (2010/670/EU). *OJ*, L 290/39, 06 November, 2010.

Commission. (2011a). *Green Paper: From Challenges to Opportunities: Towards a Common Strategic Framework for EU Research and Innovation Funding.* COM (2011) 48 Final, 9 February. Brussels: European Commission.

Commission. (2011b). *'Analysis of Public Consultation': Green Paper on a Common Strategic Framework for EU Research and Innovation Funding, 2011* (Undated). Brussels: European Commission.

Commission. (2011c). *Proposal for a Council Decision Establishing the Specific Programme Implementing Horizon 2020 – The Framework Programme for Research and Innovation (2014–2020).* COM (2011) 811 Final 2011/0402 (CNS), 30 November. Brussels: European Commission.

Commission. (2011d). *Proposal for a Council Decision Concerning the Framework Programme of the European Atomic Energy Community for Nuclear Research and Training Activities (2012–2013).* COM (2011) 72 Final, 7 March. Brussels: European Commission.

Commission. (2013a). *Energy Technologies and Innovation.* COM (2013) 253 Final, 2 May. Brussels: European Commission.

Commission. (2013b). *Review of the SET-Plan Implementation Mechanisms for the Period 2010–2012.* Undated. https://setis.ec.europa.eu/system/files/SET-Plan_%20Review%20of%20Implementation%202010-12.pdf. Accessed 02 Oct 2018.

Commission. (2014a). *The ERA-NET Scheme from FP6 to Horizon 2020* (Report on ERA-NETs, Their Calls and the Experiences from the First Calls Under Horizon 2020). https://www.era-learn.eu/documents/ec-publications/the-era-net-scheme-from-fp6-to-horizon-2020.pdf. Accessed 02 Oct 2018.

Commission. (2014b). *Strategic Energy Technology (SET) Plan: Towards an Integrated Roadmap: Research & Innovation Challenges and Needs of the EU Energy System*, JRC93056, Version December 2014. https://setis.ec.europa.eu/system/files/Towards%20an%20Integrated%20Roadmap_0.pdf. Accessed 02 Oct 2018.

Commission. (2014c). *European Energy Security Strategy.* COM (2014) 330 Final, 28 May. Brussels: European Commission.

Commission. (2015a). *Energy Union Package – A Framework Strategy for a Resilient Energy Union with a Forward-Looking Climate Change Policy.* COM (2015) 80 Final, 25 February. Brussels: European Commission.

Commission. (2015b). *Towards an Integrated Strategic Energy Technology (SET) Plan: Accelerating the European Energy System Transformation*. C (2015) 6317 Final, 15 September. Brussels: European Commission.

Commission. (2015c). *Research Fund for Coal and Steel – About*. Webpages Updated 25 November 2015. http://ec.europa.eu/research/industrial_technologies/rfcs_about.html. Accessed 25 June 2018.

Commission. (2016, September). *Innovative Financial Instruments for First-of-a-Kind, Commercial-Scale Demonstration Projects in the Field of Energy*. Report Written by ICF in Association with London Economics. Brussels: European Commission.

Commission. (2017a). *The Strategic Energy Technology (SET) Plan, at the Heart of Energy Research and Innovation in Europe, 2007–2017, SET-Plan 10th Anniversary*. Luxembourg: Publications Office of the European Union.

Commission. (2017b). *Energy R&I Financing and Patenting Trends in the EU, Country Dashboards 2017 Edition* (JRC Science for Policy Report, EUR 29003 EN). Luxembourg: Publications Office of the European Union.

Commission. (2017c, December 20). *Turning Solar and Bioenergy into Electricity: Two New Technologies Will Enter Commercial Demonstration Phase Thanks to EU Loans*. Brussels. http://ec.europa.eu/research/index.cfm?pg=newsalert&year=2017&na=na-201217. Accessed 16 Aug 2018.

Commission. (2017d). *EURATOM Success Stories in Facilitating Pan-European Collaborative Research Activities*. Presentation by DG Research at 10th Annual International Nuclear Conference, Piteşti, Romania, 24–26 May 2017. https://www.nuclear.ro/conference/nuclear%202017/prezentari_sesiuni_plenare/Roger%20Garbil-EURATOM%20success%20stories.pdf. Accessed 16 Aug 2018.

Commission. (2017e). *Interim Evaluation of Indirect Actions of the EURATOM Research and Training Programme*. Commission Staff Working Document SWD (2017) 427 Final, 01 December. Brussels: European Commission.

Commission. (2018a). *Report on the Implementation of the European Energy Programme for Recovery and the European Energy Efficiency Fund*. Brussels. COM (2018) 86 Final, 05 March. Brussels: European Commission.

Commission. (2018b). *Smart City Lighthouse Projects*. https://eu-smartcities.eu/group/454/description. Accessed 20 June 2018.

Commission. (2018c). *Horizon 2020 Advisory Group on Energy (E02981)*. Register of Commission Expert Groups and Other Similar Entities. http://ec.europa.eu/transparency/regexpert/index.cfm?do=groupDetail.groupDetail&groupID=2981. Accessed 20 Aug 2018.

Council. (1983a). Council Resolution of 25 July 1983 on Framework Programmes for Community Research, Development and Demonstration Activities and a First Framework Programme 1984 to 1987. *Official Journal*, 4.8.83, No C 208/1.

Council. (1983b). Decision of 22 December 1983 Adopting a Research Programme to Be Implemented by the Joint Research Centre for the European Atomic Energy Community and for the European Economic Community (1984 to 1987), (84/1/EURATOM, EEC), 5. 1.84. *Official Journal of the European Communities*, L 3/21.

Council. (2008). *Council Conclusions on a European Strategic Energy Technology Plan*. 2845th Transport, Telecommunications and Energy Council Meeting. Brussels, 28 February.

Council. (2011). Decision of 19 December 2011 Concerning the Framework Programme of the European Atomic Energy Community for Nuclear Research and Training Activities (2012 to 2013). *OJ*, L47/25, 18 February, 2012.

Council. (2013a, June 27). *Presidency Note on Horizon 2020*. Brussels. http://data.consilium.europa.eu/doc/document/ST-11627-2013-INIT/en/pdf. Accessed 14 June 2018.

Council. (2013b). Main Results of the Council, 3243rd Council Meeting, *Transport, Telecommunications and Energy*, Luxembourg, 6, 7 and 10 June 2013. Press Release. http://www.consilium.europa.eu/uedocs/cms_data/docs/pressdata/en/trans/137408.pdf. Accessed 15 June 2018.

EERA. (2010, October). *FP8 Position Paper of the European Energy Research Alliance (EERA)*. http://ec.europa.eu/research/horizon2020/pdf/contributions/prior/eera.pdf. Accessed 13 June 2018.

Eikeland, P. O. (2012). *EU Energy Policy Integration: Stakeholders, Institutions and Issue-Linking* (FNI Report 13/2012). Lysaker: Fridtjof Nansen Institute.

Energy Post. (2013, July 31). Is the EU Done Picking Clean Energy Winners? http://energypost.eu/is-the-eu-done-picking-clean-energy-winners/. Accessed 14 June 2018.

ESTI. (2016). *European Solar Test Installation*. https://ec.europa.eu/jrc/en/research-facility/european-solar-test-installation. Accessed 15 Mar 2017.

Euractiv. (2005, October 31). Blair Calls for Stronger EU Energy Policy Co-operation. https://www.euractiv.com/section/science-policymaking/news/blair-calls-for-stronger-eu-energy-policy-co-operation/. Accessed 11 June 2018.

Eurobarometer. (2008, September). *Europeans' Attitudes Towards Climate Change* (Report). Brussels: European Commission.

European Council. (2000). *Lisbon European Council 23 and 24 March 2000, Presidency Conclusions*. http://aei.pitt.edu/43340/1/Lisbon_1999.pdf. Accessed 08 June 2018.

European Council. (2002). *Barcelona European Council 15 and 16 March 2002*. http://ec.europa.eu/invest-in-research/pdf/download_en/barcelona_european_council.pdf. Accessed 09 June 2018.

European Council. (2003). *Brussels European Council 20 and 21 March 2003*, Presidency Conclusions. https://www.consilium.europa.eu/media/20858/75136.pdf. Accessed 09 June 2018.

European Council. (2006). *Brussels European Council 23/24 March 2006*, Presidency Conclusions. http://www.consilium.europa.eu/ueDocs/cms_Data/docs/pressData/en/ec/89013.pdf. Accessed 10 June 2018.

European Council. (2007). *Presidency Conclusions from European Council 8 and 9 March 2007*. Brussels. https://www.consilium.europa.eu/ueDocs/cms_Data/docs/pressData/en/ec/93135.pdf. Accessed 07 Dec 2018.

European Council. (2008). *Presidency Conclusions from European Council 13 and 14 March 2008*. Brussels. https://www.consilium.europa.eu/ueDocs/cms_Data/docs/pressData/en/ec/99410.pdf. Accessed 07 Dec 2018.

European Environment Agency. (2017). *Primary Energy Consumption by Fuel*. Last Modified 05 January 2017, Accessed 02 Oct 2018 at https://www.eea.europa.eu/data-and-maps/indicators/primary-energy-consumption-by-fuel-6/assessment-1

European Investment Bank. (2007, September 27). *Impressive Start for New EC-EIB Financial Instrument: Risk Sharing Finance Facility (RSFF) Contributes EUR 359 Million to Research and Innovation, with Strong Focus on Renewable Energy Technologies*. Press Release. http://www.eib.org/infocentre/press/releases/all/2007/2007-095-risk-sharing-finance-facility-rsff-contributes-eur-359-million-to-research-and-innovation-with-strong-focus-on-renewable-energy-technologies.htm

European Investment Bank. (2013, June). *Operations Evaluation: Second Evaluation of the Risk Sharing Finance Facility (RSFF) Final Report*. Luxembourg: European Investment Bank.

European Investment Bank. (2016, July 06). *EU Support for Development of Commercial Wave Energy Technology in Europe*. Press Release. http://www.eib.org/en/infocentre/press/releases/all/2016/2016-165-eu-support-for-development-of-commercial-wave-energy-technology-in-europe.htm. Accessed 02 Oct 2018.

European Parliament. (2008a). *Draft Report on the European Strategic Energy Technology Plan*. Committee on Industry, Research and Energy. Brussels, 17 March.

European Parliament. (2008b). *European Parliament Resolution of 9 July 2008 on the European Strategic Energy Technology Plan* (2008/2005(INI)). Strasbourg: European Parliament.

European Parliament. (2010). Report on the Proposal for a Regulation of the European Parliament and of the Council Amending Regulation (EC) No 663/2009 Establishing a Programme to Aid Economic Recovery by Granting Community Financial Assistance to Projects in the Field of Energy (COM(2010)0283 – C7–0139/2010–2010/0150(COD)), Committee on Industry, Research and Energy, Rapporteur: Kathleen Van Brempt.

European Parliament. (2011). *European Parliament Resolution of 27 September 2011 on the Green Paper: From Challenges to Opportunities: Towards a Common Strategic Framework for EU Research and Innovation Funding.* Strasbourg: European Parliament.

European Parliament. (2013a). *Horizon 2020: Boosting Research and Innovation* (Briefing from European Parliamentary Research Service 14 November). http://www.europarl.europa.eu/RegData/bibliotheque/briefing/2013/130661/LDM_BRI(2013)130661_REV1_EN.pdf. Accessed 14 June 2018.

European Parliament. (2013b). *Report on Current Challenges and Opportunities for Renewable Energy in the European Internal Energy Market 2012/2259(INI), A7–0135/2013.* Committee on Industry, Research and Energy. Rapporteur: Herbert Reul, 28 March. http://www.europarl.europa.eu/sides/getDoc.do?pubRef=-//EP//NONSGML+REPORT+A7-2013-0135+0+DOC+PDF+V0//EN. Accessed 15 June 2018.

European Parliament. (2016). *Research in the European Treaties*, Briefing, March 2016. Brussels: European Parliament.

European Parliament and the Council. (2002). Decision No 1513/2002/EC of the European Parliament and of the Council of 27 June 2002 Concerning the Sixth Framework Programme of the European Community for Research, Technological Development and Demonstration Activities, Contributing to the Creation of the European Research Area and to Innovation (2002 to 2006).

European Parliament and the Council. (2006). Decision No 1982/2006/EC of 18 December 2006 Concerning the Seventh Framework Programme of the European Community for Research, Technological Development and Demonstration Activities (2007–2013). *OJ* L 412/1, 30 December, 2006.

European Parliament and the Council. (2009). Regulation (EC) No 663/2009 of the European Parliament and of the Council 13 July 2009 Establishing a Programme to Aid Economic Recovery by Granting Community Financial Assistance to Projects in the Field of Energy. *OJ* L200/31, 31 July, 2009.

European Parliament and the Council. (2010). Regulation (EU) No 1233/2010 of 15 December 2010 Amending Regulation (EC) No 663/2009 Establishing a Programme to Aid Economic Recovery by Granting Community Financial Assistance to Projects in the Field of Energy. *OJ* L346/5, 30 December, 2010.

Gornitzka, Å., & Metz, J. (2014). European Institution Building Under Inhospitable Conditions: The Unlikely Establishment of the European Institute of Innovation and Technology. In M.-H. Chou & Å. Gornitzka (Eds.), *Building the Knowledge Economy in Europe: New Constellations in European Research and Higher Education Governance* (pp. 111–130). Cheltenham: Edward Elgar. isbn:978 1 78254 528 6.

Initiative for Global Leadership in Concentrated Solar Power. (2017, November). *Implementation Plan.* https://setis.ec.europa.eu/system/files/set_plan_-_csp_initiative_implementation_plan.pdf. Accessed 20 June 2018.

InnoEnergy. (2018). *Key Facts*. http://www.innoenergy.com/about-innoenergy/key-facts/. Accessed 02 Oct 2018.

IPCC. (2007). *Climate Change 2007: Synthesis Report*. Geneva: IPCC.

JRC. (2015). *Capacity Mapping: R&D Investment in SET-Plan Technologies – Reference Year 2011* (JRC Science and Policy Report, JRC95364). Luxembourg: European Union.

Markham, S. K., Ward, S. J., Aiman-Smith, L., & Kingon, A. I. (2010). The Valley of Death as Context for Role Theory in Product Innovation. *Journal of Product Innovation Management, 27*(3), 402–417.

Research Council of Norway. (2011). *Kort om SET-Planen (SET-Plan Brief)*. Internal Note, March 2011 – revised 2013. https://www.forskningsradet.no/servlet/Satellite?blobcol=urldata&blobheader=application%2Fpdf&blobheadername1=Content-Disposition%3A&blobheadervalue1=+attachment%3B+filename%3DKortomSET-planen.pdf&blobkey=id&blobtable=MungoBlobs&blobwhere=1274502878282&ssbinary=true. Accessed 12 June 2018.

Ricardo Energy & Environment. (2017). *Report on the First Results of H2020 Projects on Energy Efficiency and System Integration*. Final Report for DG ENERGY, Developed in Co-operation with CE Delft and DNV-GL, ED 62228, Issue Number V1.5, 29 March. https://ec.europa.eu/energy/sites/ener/files/documents/ed62228_h2020_energy_evaluation_final_report_v1.5_3_0.pdf. Accessed 13 June 2018.

SETIS. (2015). *Launch of InnovFin Energy Demonstration Projects (EDP) Facility to Support Energy Innovators*. https://setis.ec.europa.eu/newsroom/news/launch-of-innovfin-energy-demonstration-projects-edp-facility-support-energy. Accessed 14 June 2018.

SETIS. (2018). *Implementation Plans*. https://setis.ec.europa.eu/actions-towards-implementing-integrated-set-plan/implementation-plans. Accessed 22 Oct 2018.

SETIS. (Undated). *European Initiative on Smart Cities*. https://setis.ec.europa.eu/set-plan-implementation/technology-roadmaps/european-initiative-smart-cities. Accessed 15 Aug 2018.

SETIS Magazine. (2017a, April). Horizon 2020 ERA-NETs in the SET-Plan: The Experience to Date. https://setis.ec.europa.eu/setis-reports/setis-magazine/funding-low-carbon-technologies/horizon-2020-era-nets-set-plan. Accessed 14 June 2018.

SETIS Magazine. (2017b, November). Looking Back at 10 Years of Forward Thinking, the SET-Plan – We Need a Place Now More than Ever for Concrete Collaboration Between Countries. https://setis.ec.europa.eu/setis-reports/setis-magazine/looking-back-10-years-of-forward-thinking-set-plan/we-need-place-now. Accessed 14 June 2018.

Skjærseth, J. B. (1993). *The Climate Policy of the EC: A Study of Interests and Preferences Versus EC Problem-Solving Capacity* (FNI Report 2/1993). Lysaker: Fridtjof Nansen Institute.

Skjærseth, J. B. (1994). The Climate Policy of the EC: Too Hot to Handle. *Journal of Common Market Studies, 32*(1), 25–45.

Skjærseth, J. B. (2014). *Implementing EU Climate and Energy Policies in Poland: From Europeanization to Polonization?* (FNI Report 8). The Fridtjof Nansens Institute.

Skjærseth, J. B., Eikeland, P. O., Gulbrandsen, L. H., & Jevnaker, T. (2016). *Linking EU Climate and Energy Policies: Policymaking, Implementation and Reform.* Cheltenham: Edward Elgar.

SNETP. (Undated). *Sustainable Nuclear Power European Technology Platform Website Presentation of the Nuclear Power European Industrial Initiative.* http://www.snetp.eu/esnii/. Accessed 20 June 2018.

Turmes, C. (2017). *Energy Transformation – An Opportunity for Europe.* London: Biteback Publishing.

van der Veen, G., Altmann, M., Eparvier, P., Ploeg, M., & Trucco, P. (2014, June 19). *Evaluation of the Impact of Projects Funded Under the 6th and 7th EU Framework Programme for RD&D in the Area of Non-nuclear Energy* (Final Report, FP6/7 Energy Projects, Version 4). Brussels: Technopolis Group.

Wejnen, M., Greer, H., Forlani, F., Wagner, U., & Salminen, P. (2000). *Five-Year Assessment Report Related to the Specific Programme: Energy, Environment and Sustainable Development Covering the Period 1995–1999.* Brussels: Commission. Retrieved 19 September, 2017, from https://ec.europa.eu/research/evalua-tions/pdf/archive/five_year_assessments/five-year_assessment_1995-1999/fp5_panels_final_report_energy_environment_and_sustainable_develop-ment_2000.pdf#view=fit&pagemode=none

CHAPTER 4

Explaining Making of the SET-Plan

Why did the SET-Plan emerge, and with the six technology areas as priorities? Why was an essentially decentralized governance system established? Here we seek explanations in the role of the actors involved in making the Plan—member-states, EU institutions and non-state actors—and discuss international technology markets as contextual factors.

4.1 ROLE OF MEMBER-STATES

Although the EU member-states did not request the SET-Plan, they laid the foundations by signalling greater willingness to transfer competence in R&I and energy policy to the EU level. In 2000, the Lisbon European Council adopted the European Research Area, aimed at fighting fragmentation among national R&D policies by joint action at EU level. Then, the Hampton Court Summit of EU leaders in 2005 signalled new preferences for strengthening energy policy at the EU level. Statements made by the host of this summit, British PM Tony Blair, are illustrative: 'I believe it is time we developed within Europe a common European energy policy. For far too long we have been in the situation where, in a haphazard and random way, energy needs and energy priorities are simply determined in each country according to its needs' (European Parliament 2005). The statement signalled a significant change in attitude from the UK government, which had been among the strongest opponents of transferring to Brussels powers in energy policy, ever since joining the Community in

© The Author(s) 2020   63
P. O. Eikeland, J. B. Skjærseth, *The Politics of Low-Carbon Innovation*, https://doi.org/10.1007/978-3-030-17913-7_4

1973.[1] Declining oil and gas resources had brought the United Kingdom more in line with other EU countries, increasingly dependent on imports and more concerned about energy security. The United Kingdom thus recognized stronger EU energy policy integration as compatible with its national interests.

Further traces of stronger member-state willingness for integration are found in responses to the Commission 2006 Green Paper which spelled out possible options for a future EU energy policy—including a new SET-Plan (Eikeland 2012). The UK government called for maintaining the momentum in EU energy policy built up at Hampton Court and welcomed greater integration of energy policy to encompass also a SET-Plan with strong focus on energy-efficient and clean-energy technologies. Similar responses came from other major EU member-state governments. Both the German and French governments supported the idea of a SET-Plan, emphasizing the development of CCS in industry as an important priority. France also stressed that EU policies should focus on promoting zero-carbon energy sources broadly, including nuclear power, and not only on renewable energy.

Most of the new Central and Eastern European Countries (CEECs) held that security of supply should be given highest priority in future EU energy policy, but were generally silent on the idea of an EU SET-Plan, indicating indifference. Their views reflected the wish to become less dependent on natural gas supplies from Russia. Fears of their big neighbour were fanned by conflicts between Russia and Ukraine in early 2006 that had cut off supply during cold winter days. Many CEECs urged for an EU energy policy that could focus on alternative channels of gas supply, better East–West energy-system interconnection, and policies to keep open any energy-mix option that would favour the use of domestic energy sources, including coal (Eikeland 2012).

That the member-states signalled willingness to coordinate R&D and energy policies at EU level constituted a window of opportunity for the Commission to propose and develop the SET-Plan idea. The Commission got broad backing for new EU-level energy policies: the SET-Plan idea was endorsed by the major member-states (Commission 2006), with no direct opposition expressed. The Plan could thus be launched as a part of a comprehensive new energy strategy with a focus on indigenous low-carbon energy sources.

---

[1] Apart from creating a level playing field for the internal energy market.

Despite these new integration-supportive policy signals, the member-states failed to accept the transfer of much authority to EU institutions when discussing how the proposed SET-Plan should be governed. The Commission contemplated and assessed four different governance options (Commission 2007d). One envisioned a pure market-based selection of technologies based on technology-neutral market deployment policies—not in line with the fundamental idea that the SET-Plan would select priorities. The other options would designate various competences between the member-states and the EU level for selecting technologies and steering resources for implementation (Table 4.1).

One of these was a top-down EU-centralized approach—which was assessed as politically unfeasible, as the EU Treaty settled the principle of 'shared competence' for R&I policies. The Commission thus recommended that a mixed bottom-up/top-down approach be established, under an appropriate legal basis or under the Open Method of Coordination (OMC). A legal basis could imply the adoption of enforcement mechanisms; by contrast, OMC would only provide for soft governance mechanisms (European Parliament 2014).

However, the Council provided no new legal mechanisms for the EU level to enforce effective implementation of the SET-Plan. On the contrary, the Council reinforced control by not giving the SET-Plan any legal status or other formal authority. The Plan was not finalized in a binding text summarizing a compromise between unsettled disagreements on direction, priorities and governance. The Council assigned governance of

Table 4.1 Policy options for governing the SET-Plan

Policy options for governance	Levels of activities	Sharing of competence
Bottom-up/ status quo	Individual programmes at EU, national and corporate levels	Member-states' programmes driving force. Weak coordination with EU programmes
Mixed	Strategic coordinated framework including EU, national and corporate levels	Between member-states with authority over budget and the commission chairing a coordination committee
Top-down/ centralized	Centralized holistic structure at EU level	Strong EU institution competence through a governing board based on funding by the member-states, the commission and industry

Source: Based on Commission (2007d)

the Steering Group to the member-states; it also stressed that the Plan should have no implications for financial means, thus restricting the Commission's room for manoeuvre (see Chap. 3). The governance model chosen reflected member-state preferences for retaining control over strategic technologies and R&I resources. This model had clear weaknesses, as the Commission acknowledged. Potential discrepancies between the strategic priorities of the Plan and long-term stakeholder commitment might challenge its future funding (Commission 2007d).

What role did member-state preferences play for the Commission's selection of priority technologies for the SET-Plan? A strong role for the member-states would be indicated by a high match between priorities in national energy R&I funding with those selected for the Plan. National priorities could be expected to reflect national strategic-energy interests shaped by the availability of primary energy resources, existing energy-system technologies and the priorities of national energy-technology manufacturers.

Before the SET-Plan idea was launched, the EU member-states varied significantly in their levels of spending and priorities for public energy research. In 2005, nearly half of the member-states had no special national or regional energy R&D programmes (JRC 2008). France, Germany and Italy alone accounted for 73 per cent of total EU spending on energy research (above €350 million/year each). Five additional countries (Finland, the Netherlands, Spain, Sweden and the United Kingdom) contributed substantially, spending between €50 million and €150 million/year (JRC 2008, p. 40). The new CEECs—many of them highly coal-dependent—accounted for less than 3 per cent of total EU spending, resulting in the observed indifference to the SET-Plan idea. Heterogeneous R&D infrastructures were paralleled by great national variation in market-pull policies for new technologies, with feed-in tariffs and tradable certificates being the most prominent (JRC 2008, p. 59).

Trends in priorities for national public R&D among the major funding countries aligned roughly with what the Commission selected as initial SET-Plan priorities (JRC 2008).[2] Renewable energy technologies were the most important priorities, mainly renewable electricity from wind, solar, and bioenergy. Windpower was already a relatively mature industry in some countries. Three-quarters of total EU R&D on windpower

---

[2] Public energy research figures are surrounded by high uncertainty but are considered sufficiently reliable to show general trends.

2000–2005 had been conducted in Denmark, Germany and the Netherlands, and had been gaining in priority in France.

Photovoltaic solar technology (PV) had recently experienced very high growth rates (annual average of more than 30 per cent) in some countries like Germany and the Netherlands. Economies of scale in production had reduced prices by a factor of five over the past 20 years, with increased efficiency and systems reliability. The sector was characterized by small- and medium-sized enterprises (SMEs) highly dependent on public R&D subsidies. High Temperature Solar Thermal Technologies were strongly and continuously supported by Germany and Spain; and Italy had been stepping up its R&D efforts.

Bioenergy had high priority in many countries, with the highest or second highest share in total energy research in Austria, Denmark, Hungary, the Netherlands and Sweden. Public R&D spending was on the increase also in Germany. The EU figured among the world leaders in the use of biomass for electricity generation. For Bulgaria, Poland, Romania, biomass was considered a strategic asset, making bioenergy an opportunity to attract also the interest of new member-states. Finland, the Netherlands and Sweden reported giving top priority to biofuels for the transport sector.

France was the major funder of R&D on nuclear power, spending over 60 per cent of total national energy R&D on this priority. Other member-states recorded shares in the range of 0–30 per cent. One out of three member-states listed nuclear related research among the top priorities. This included those with relatively large nuclear-power sectors, like Bulgaria, the Czech Republic, France and Lithuania.

By contrast, CCS technology development had received rather limited funding at EU and national levels, and with a lack of cooperation recorded between industry and governments. However, CCS had become a more recent priority, not yet reflected in figures on R&D spending, except for the United Kingdom, where it accounted for 10 per cent of total public energy R&D. CCS ranked high on the R&D agenda also in France, Italy and Germany.[3] That the SET-Plan included CCS despite its relatively low priority in national research underscores its significance for dampening political opposition to EU climate and energy policy among fossil-fuel interests.

---

[3] R&D budgets were around €10–20 million per year in France, Germany and the United Kingdom.

What about the technology areas that were not accorded priority in the SET-Plan—did these have low priority also at the national level? That may well be. Member-states did not identify hydrogen and fuel cells as a top priority, except for Denmark, which invested more than a quarter of its total energy R&D in hydrogen technologies.[4] Geothermal energy was similarly not a high-priority R&D-area for most countries, and ocean energy for power generation (wave and tidal) had only small budgetary shares in Denmark, Ireland, Portugal, Sweden and the United Kingdom.

Puzzling exceptions to the pattern included some technology areas given high priority in the member-states but not by the SET-Plan. Energy efficiency ranked very high in many member-state R&D budgets, with spending on the increase in France and Germany. There was a strong focus on improving the efficiency of fossil-fuel converting technologies such as gas turbines, coal gasification systems and clean-coal combustion technologies. Member-states with major budget shares for such research included France, Germany, Italy, the Netherlands and the United Kingdom. As these countries were also among the climate-policy 'pushers', they were willing to accept a low-carbon technology focus for the SET-Plan (Skjærseth et al. 2016).

Thus, public R&D priorities and interests among the major member-states corresponded largely, but not fully, with the Commission's initial selection of SET-Plan technologies for EIIs: wind, solar, bioenergy, CCS, electricity grid and nuclear fission. The Commission excluded some technologies that were prioritized by the member-states, most notably energy-efficiency solutions for several sectors, including clean-coal technologies. Such technologies had been specifically emphasized by some of the CEECs, notably Poland and the Czech Republic, in upfront consultations for the SET-Plan (Chap. 3). Thus, we see that the Commission made some real priorities that ran counter to the R&D interests and priorities of several member-states.

To sum up, the SET-Plan was not requested by the member-states. However, temporary changes in the positions of member-state governments towards stronger integration in EU research, climate and energy policies created a window of opportunity for the Commission to propose and develop the Plan. The governance system and the initial selection of technologies largely mirrored member-state R&I interests and priorities and their preferences for a decentralized bottom-up approach.

---

[4] The relative share of government funding in this area remained below 6 per cent for other member-states.

4.2   Role of EU Institutions and Non-State Actors

4.2.1   The Commission

The Commission initiated the SET-Plan: prior to launching the idea, it also played important agenda-setting roles. Commission services first provided the diagnosis that led the European Council in 2000 to adopt the European Research Area: inadequate and poorly coordinated EU R&I efforts had prevented European companies from retaining and winning shares in global technology markets, thus putting Europe on a trend of deteriorating international competitiveness (Commission 2000). The Commission-appointed Advisory Group on Energy, tasked with assisting in developing the energy portions of the FP, documented similar weaknesses regarding EU energy-technology R&I. The group suggested revisions: EU energy R&D programmes should be strengthened, have a more specific strategic focus, and research and demonstration components should be better integrated (Wejnen et al. 2000, p. x; ERAWOG 2005).

Poor coordination at the EU level reflected the persisting cooperation problems between the Commission services that had been designated lead roles in managing EU funds for energy R&D and demonstration projects—DG Research and DG Energy, respectively (Wejnen et al. 2000, p. ix; Advisory Group on Energy 2006). Tensions between the two reflected their different roles in EU policymaking, which also led to varying perceptions on how the EU R&I budget should best be spent. DG Research would have lead responsibility in drafting EU Framework Programmes and the main role in promoting and securing funding for top-level research in European research institutions, for the long-term benefit of all sectors in society. DG Energy would have lead responsibility for FP funds allocated to demonstration and deployment programmes, especially as regards ensuring that such programmes would respond to shorter-term energy policy objectives and energy-industrial challenges.

Notwithstanding the long-term tensions, policy officers in DG Research and DG Energy teamed up to draft and develop the SET-Plan idea as a common project. On the team were also representatives of a third Commission service, the Joint Research Centre (JRC)—the Commission's own research-executing branch, which had also been assigned a central role to assist the Commission developing policies where scientific and technological assessments would be needed (Commission 2017). The trio responded to new energy-policy integration signals from the member-states and assessments indicating that a greater strategic approach to EU energy research would be needed to accelerate innovation and market uptake of new low-carbon technologies. They started by discussing how their different interests could be combined and how the Plan could fill a void and thereby create added value (Interview B). The key DG Energy officer on the team explained how the SET-Plan represented an opportunity to bridge differing EU energy and general research interests: the idea made 'common sense' since the Framework Programme for Research was then spreading its energy budget 'thinly across many small projects, involving many partners and pursuing multiple technologies, with no priorities or targets and without monitoring results' *(SETIS Magazine,* November 2017).

For DG Energy, the SET-Plan represented an opportunity for strengthening and better steering energy R&I programmes towards solving the EU energy-system challenges to which it was tasked to respond, related to energy security, climate change and competitiveness. A new and stronger political focus on energy R&I could contribute to halting the long-term decline in EU energy R&I programmes, which was paralleled by a long-term faltering engagement in research and demonstration of new technologies among leading energy utilities and energy-technology companies (Eikeland 2013; Interview B). Utilities and technology manufacturing firms were poorly committed to projects conducted at EU level: the Plan offered an opening to increase their commitment and break the trend of industrial R&I capacities moving out of Europe (Interview B).

For DG Research, the SET-Plan provided a new opportunity for getting the research community to proceed with implementation of the ERA by tighter clustering of the research community and industry (Chap. 3). The ERA component of the SET-Plan (European Energy Research Area— EERA) thus fits well with the main mission of DG Research, and the EII component, to boost industrial demonstration efforts, fits well with DG Energy tasks (Interview G).

For the JRC, the Plan represented an opportunity to upload its own research interests and retain its role as a leading EU energy R&I agent in selected areas of technology, like nuclear and solar power. Concerning nuclear power, new research and coordination tasks had recently been assigned to the JRC by connecting EURATOM to the Generation IV

International Framework Agreement (Commission 2005a, p. 11).[5] To advance the specific SET-Plan idea, Commission staff in the JRC, DG Research and DG Energy focused on common interests and managed to convince their respective directors to keep the idea on the policy agenda by drafting the first 2007 SET-Plan communication (Interviews D, N and R).

The new closer cooperation between these services represented a larger trend of far more extensive Commission interservice cooperation than previously for DGs with lead responsibilities in energy, research and climate policies. Internal Commission rivalry and disagreement among DGs had delayed progress in climate and energy policymaking (Eikeland 2011; Skjærseth et al. 2016). A severe conflict between DG Energy and DG Environment had surfaced following the adoption of the EU ETS and EU ratification of the Kyoto Protocol in 2002. Energy Commissioner de Palacio had challenged Environment Commissioner Wallström over the economic costs of the latter's climate leadership-by-example ambition, questioning the rationale, as well as the costs that would be involved in adopting measures to implement the EU's Kyoto commitment if the Protocol failed to enter into force. Commission President Romano Prodi had then publicly criticized de Palacio and stressed the importance of keeping the Commission unified in support of the EU's leadership ambition in international climate policies (EurActiv 2005a; Barnes 2011).[6]

The key to unlocking the differences in interests between DG Research, DG Energy and DG Environment was to develop an integrated package of policies with elements that were mutually dependent and could offer something to everyone (Skjærseth et al. 2016). In January 2007, the Commission issued the three communications—on energy, technology and climate policies for 2020 and beyond (Commission 2007a, b, c). These set the tone for a shared, radical project: to transform Europe by speeding up the transition to low-carbon growth, in order to set the pace for a new global industrial revolution (Commission 2007b, pp. 5, 21).

[5] JRC's energy research interests were wider and included a new fuel cells and hydrogen (FCH) test facility set up in 2005 (Commission 2005a, p. 9). JRC was instrumental in establishing FCH as the first ETP and giving this status as a JTI.

[6] Concerns about costs were shared by the 'competitiveness-first' DGs responsible for the internal market and industry/enterprise. Those critical to the costs of environmental policies and the impact on competitiveness of European industry included Enterprise Commissioner Verheugen and Internal Market Commissioner McCreevy (EurActiv 2005a, b).

The specific mix of policies proposed was framed so as to ensure that all major energy, climate and competitiveness concerns would be addressed. The Commission put pressure on the decision makers to accept *all* the policies proposed, including the SET-Plan, to meet the 20-20-20 goals. Synergies between climate, technology and energy policies were underscored. Action on climate change was placed at the centre of a new EU energy policy by stressing the importance of making energy use more efficient, lessening the need for imported hydrocarbons and reducing vulnerability to fluctuations in oil and gas prices. Action on energy policy would contribute to climate-change mitigation and more effective application of the ETS, while also creating new 'green' jobs. All this was to be achieved by strengthening policies on renewables, energy efficiency, liberalization of the European energy market and innovation in low-carbon technologies.

In March 2007, the Commission's SET-Plan idea finally succeeded: the Plan received backing from European Council as an integral element in the new policy package. In November 2007, initiation of the Plan was completed with the Commission's final SET-Plan communication. The new arenas for governing implementation of the Plan were based largely on previous Commission initiatives. The basis for EERA were various networking activities among European research institutions tied to implementing the ERA, while the EIIs built on the European Technology Platforms that served as network arenas for clustering industrial actors and research institutions with the Commission as coordinator. The proposed SETIS information system had been elaborated by the JRC, originating from a project that started in 2000 and became the Assessment of Clean Energy Technologies project in 2003. An energy-technology monitoring system was created to inform the JRC and the Commission in pursuit of a sustainable energy policy (Commission 2012, p. 93). JRC staff engaged in this project became part of the joint team, with DG Energy and DG Research, that engaged in the conception and development of the SET-Plan, including the information system named SETIS.

Thus, the Commission and its Joint Research Centre prepared the ground, launched the idea and initiated the SET-Plan as an integral element in a wider climate and energy package. The Commission proposed a governance system that partly reflected how R&I policies were already defined as a shared competence and that, with a proper legal base, could have extended competence to control implementation at EU level.

As to priorities, the Commission, through its proposal for European Industrial Initiatives, selected the technology areas and thereby the actors who would benefit most from the Plan. The Commission narrowed in on what was essentially a portfolio of electricity-supply technology areas (generation and transmission/distribution), with biofuels for the transport system as an exception, and emphasized that the Plan should consider supporting especially companies in the energy sector, notably electricity-supply companies (Commission 2007d, p. 11). The justification was the need for government funding to compensate low R&D intensity stemming from the market liberalization that had intensified price competition and reduced monopolistic rents for the companies.

The six technology areas were selected from a far wider range of possible low-carbon electricity technologies at various stages of development. Several criteria were employed, including EU added value/ additionality and industry and member-state willingness to join forces (see Chap. 3). The EU added-value criterion was operationalized by the Commission as large-scale projects that required resources beyond the capacities of individual member-states. Several technology areas (like CCS, nuclear power and biorefineries) would by their very nature require large investments to realize full-scale demonstration, whereas others could be demonstrated also on a smaller scale, like solar power (Commission 2016, p. 10).

The 'willingness to join forces' criterion was operationalized by the Commission as referring mainly to technology areas covered by the European Technology Platforms, networks of industrial companies and associations, researchers, national governments and other stakeholders that, on the basis of earlier collaborative research projects in FP6, had already kick-started developing strategic research agendas for selected areas of technology.

The Commission's selection of technology areas for the SET-Plan proposal was very much the result of a bottom-up process where national priorities and alliances of non-governmental and governmental actors that proved most effective in advancing joint strategic planning would also be given precedence in shaping the priorities of the Plan. However, the Commission was not a passive agent: it served as facilitator of the ETPs, thus giving these technology areas a higher status and more coordination assistance at EU level than those lacking the ETP label.

While industry would play a leading and initiating role of ETPs, the Commission would have key roles in mobilizing actors by taking a role in ETP steering structures (Commission 2004). This provided the selection of priorities via ETPs with an element of top-down steering of the process.

The Commission could justify selection by referring to existing networks for some technology areas, while also reinforcing such networks. The Commission could justify exclusion of technology areas by referring to less-mature networks. This selection procedure encountered criticism from the research community. The Commission's Advisory Group on Energy warned that the ETP label for some technologies and not others could evoke suspicions that the relative strength of lobby groups would be decisive for the allocation of energy R&I funds (Advisory Group on Energy 2006, pp. 9–11).

In addition to the ETP selection procedure, the Commission sought broader advice for the selection of priorities to the SET-Plan. This involved consultations with experts from key energy-technology sectors, hearings, online questionnaires and broad interservice consultations (Commission 2007d, p. 5). Consultations back in 2006, on the Energy Green Paper where the SET-Plan idea was launched, had received opinions from a wide range of stakeholders, including many member-state governments. 'Capacities mapping' by the JRC had identified focus of and trends in research interests and funding in the member-states and industry. The research community, represented by the Advisory Group on Energy, had provided input by a broad mapping of technology areas that could be candidates for the Plan (Commission 2007d). Included were an extensive portfolio of low-carbon electricity, heat and transport technologies, as well as a broad range of end-use energy-efficiency solutions (Commission 2007d).

Consultations showed considerable support for a SET-Plan, but diverse opinions and interests as to what technology areas to prioritize. The final selection reflected the Commission's efforts at securing continued support for the Plan by combining the diverse R&I interests of member-states and industries.

The Commission did not include all important interests with its initial proposal. For instance, energy-efficient end-use was not included, even though consultations and R&I data had shown this to be a broadly dispersed area of interest. The member-states and the European Parliament (see below) nevertheless insisted on including energy efficiency, to which the Commission responded in 2009 by adding the Smart Cities and Communities Initiative. This expanded the range of actors who would benefit from EU low-carbon R&I funds compared to what had been intended for the six initial EIIs.

4.2.2   *The European Parliament*

Although the European Parliament (EP) was not assigned a formal role in the SET-Plan governance system, it had an important role in implementation, as a co-decision maker of EU's R&I programmes. Upfront support from the EP on the Plan and its priorities could strengthen it as an EU-level energy R&I steering instrument.

The Parliament's SET-Plan involvement started from an 'own initiative' report of 2008 drafted by its Committee on Industry, Research and Energy (ITRE), with opinions from the Environment, Public Health and Food Safety Committee (ENVI). The report welcomed the Commission's SET-Plan proposal and supported the envisaged EIIs as central for attaining the 2020 targets, but also urged that energy efficiency be added as another EII (European Parliament 2008a). However, it also emphasized that the Plan should not lead to reallocation of funds already decided for FP7, or duplication of funding the EIIs from different parts of the programme.[7] Nor should funding the EIIs take proceed at the expense of research into many less-mature technologies with longer-term perspectives; this mirrored the views and interests of DG Research (Interviews D and G). The report noted auctioning revenues from the EU ETS as a funding opportunity—an issue that would reappear in later negotiations on the ETS and CCS directives (see Chap. 3). The Parliament signalled, like the Council, that funding of the SET-Plan should not be taken for granted.

Members of the ITRE committee tabled 69 proposals for amendments, a few revealing highly different opinions on the Commission's technology selection and justification for this selection. One amendment would change perceived prospects for CCS—it was proposed that the Commission wording that CCS 'will be' of critical importance in tackling climate change be altered to 'could be' of critical importance (European Parliament 2008b). The ENVI Committee, like ITRE, called for an additional EII for energy-efficiency technologies and clearer priorities among the EIIs (European Parliament 2008c). A further EII selection criterion was suggested: to consider the life-cycle of each technology and its environmental impact at each stage of the production process. This would link low-carbon technologies and climate concerns to possible trade-offs with nature conservation. Based on these inputs, ITRE submitted its motion to EP plenary voting, for adoption as final resolution (European Parliament 2008d).

---

[7] The Joint Technology Initiatives and the Innovation Framework Programme (CIP).

The final EP resolution adopted in July 2008 sent equivocal signals on the criteria for selecting technologies. It asked the Commission to consider 'replication in the longer term' and 'life-cycle impacts' as additional criteria. With foresight, the Parliament expressed doubts that CCS could play an important role towards 2020; it was also undecided as to whether EU resources should be allocated to demonstrating mature technologies for 2020 or rather for developing less-mature technologies not initially covered by the SET-Plan. This mirrored tensions within the Commission, between DG Research and DG Energy, the former taking a longer-term perspective, the latter 'not wanting results in 50 years' (Interview G).

The EP expressed strong support for the six proposed EIIs, while also requesting the Commission to investigate additional EIIs for other sectors and technologies: co-generation, hydrogen, construction and housing, a distinct focus on energy storage, heating and cooling systems, distribution infrastructures and interconnection of networks (European Parliament 2008d).[8] Like the Council, the EP insisted that energy efficiency should figure more prominently in the SET-Plan; the Commission acted on this in 2009 by proposing the 'Smart Cities' initiative.

4.2.3   Industry and Research Community

Research and industry communities were involved in the making of the SET-Plan. Broad commitment to and unity in preferences for priorities from these groups could strengthen opportunities for getting the Plan adopted and facilitate later implementation, as the research community and industry were in control of most R&I resources. Industrial companies were the principal funder of EU energy R&I and the salient actors in bringing less-mature technology pilots from the lab to the market via demonstration projects.

4.2.3.1 Industry

The period before the launch of the SET-Plan saw massive mobilization of industry (major energy companies and industry associations) around new arenas established by the Commission for discussing EU climate, energy and technology policies. From 2000, companies and industry associations had become central actors in various low-carbon technology working

---

[8] Including co-generation, hydrogen, construction and housing, energy storage, heating and cooling systems, distribution infrastructures and interconnection of networks.

groups under the European Climate Change Programme (ECCP) launched by DG Environment (Skjærseth et al. 2016). In 2005, major technology and energy companies gained seats in a new DG Enterprise and Industry-led High-Level Group on Competitiveness, Energy and the Environment aimed at promoting closer coordination between policy and legislative initiatives. The Commissioners of Energy, Environment and Competition were additional partners. One task for the group was to advice on the development and uptake of environmental and other innovative technologies. In parallel, DG Research assigned to industry a key role in the new European Technology Platforms, to develop strategic research agendas for different energy-technology areas, in line with ERA implementation guidelines. DG Energy, traditionally the key interlocutor for EU energy industries, stepped up consultation activities for its new energy-policy initiatives (Skjærseth et al. 2016).

Key industrial actors with stakes in energy-technology R&I included manufacturers and users of energy technology and their EU-level associations. Technology manufacturing companies (like GE, Siemens, Alstom, ABB and Areva) were members of Orgalime[9] and/or EPPSA.[10] Important users of technology included electric-power producers, energy-grid companies and energy-intensive industries, organized in various EU-level associations like Eurelectric,[11] Business Europe,[12] and specific energy-intensive industry branch associations. Other relevant stakeholders were the various associations established to promote their specific technologies in EU policymaking.[13]

The most important industry associations endorsed the SET-Plan idea. Orgalime favoured the development of lead markets for innovation to ensure that Europe would remain a world leader in energy technologies (Orgalime 2006). EPPSA voiced support for the Plan to focus primarily

---

[9] Orgalime is the European federation representing European mechanical, electrical, electronic and metal manufacturing industries as a whole—the largest industrial branch in the EU.

[10] EPPSA (European Power Plant Supplier Association) organizes technology companies that manufacture thermal power plants.

[11] Eurelectric organizes national electric utility associations and major utilities at EU-level.

[12] BusinessEurope is the European umbrella organization for national trade associations, representing all major business interests in Europe.

[13] These included EURACOAL, FORATOM (nuclear power), EWEA (wind power), EPIA (photovoltaics), AEBIOM (bioenergy) and EGEC (geothermal energy), COGEN-Europe (co-generation of heat and power).

on industrial demonstration (EPPSA 2007). Major technology users were also supportive: both Eurelectric and BusinessEurope welcomed more public funding for energy R&I, and acknowledged the need to correct the fragmentation in research and pool national resources in order to enable industrial demonstration projects (BusinessEurope 2007; Eurelectric 2007). This general industrial support for the SET-Plan is understandable, given the aim of the Plan to increase resources available for industrial research and demonstration projects as a key mode of accelerating innovation in energy technology. The SET-Plan represented an opportunity for higher public energy R&I funding budgets, with a larger share reserved for industry.

But the various industry associations expressed differing preferences as to the technology areas to be covered by the Plan. Orgalime, organizing companies which, collectively, had very broad interests across areas of technology, felt that the focus should be on integrated systems rather than individual technologies, and noted the close relation to the European Technology Platform 'Smart Grids'. Additionally, Orgalime emphasized that the SET-Plan should not discriminate one technology or source against another: in the interest of secure energy supply, a mix of technologies for energy production should take precedence over selecting individual ones or addressing individual topics only. Technologies highlighted as very important included smart grids, power quality and technologies for improving transnational grid stability, communication, energy efficiency (e.g. improvements of existing power plants or grids), and storage technologies (Orgalime 2006). EPPSA, deeply involved in manufacturing thermal power plants, wanted a prominent place for CCS. As a supporter of the creation of the ETP on Zero Emissions Power Plants, EPPSA sided with this ETP, pressing for ten to twelve demonstration plants by 2020 (EPPSA 2007).

Eurelectric, representing an industry with individual companies actively involved across the ETPs, and itself active on the 'Smart Grid' and Zero Emissions Platform (CCS), supported the Commission's proposal, but insisted on a wide-portfolio approach: the Plan should take into account a wide range of electricity production and end-use technologies. Eurelectric's first priority was further enhancement of the efficiency of existing or new electricity-generation technologies—whether in renewables, nuclear or fossil fuels (Eurelectric 2007). Large-scale development and deployment of CCS, new and improved nuclear technologies and a range of renewable energy technologies should all be priorities. Among new end-user technologies, particular attention should be given to heat pumps and plug-in hybrid vehicles, in line with Eurelectric's strategy of promoting the electrification of other sectors. Eurelectric warned against the increased use of bioenergy (Eurelectric 2008). Accelerating the development of biofuels would run counter to Eurelectric's strategy of promoting electrification in the transport sector.

BusinessEurope, representing energy-consuming sectors and industries broadly in Brussels, called for a strong focus on demand-side/energy-user technologies, especially those that could foster the competitiveness of energy-intensive industries (BusinessEurope 2007). Noting that such demand-side technologies had not been proposed as EII by the Commission, it voiced concern over the mode of selecting priorities, emphasizing that the market and science should drive technological development and deployment: science-led procedures should determine the choice of technologies for the Plan as independently as possible from political considerations. Procedures should include a common and consistent methodology; and technologies contributing most cost-effectively to sustainability, security and competitiveness should be regarded as having highest technological potential (BusinessEurope 2007).

Other companies and associations representing manufacturers and users of specific technologies/energy products wanted the Plan to ensure higher R&I budgets for their technologies (Interview B).[14] Some, like the natural gas and coal industries, supported the ETP on CCS and lobbied intensely to get CCS technologies included in the Plan, a precondition if their products were to have a place in a low-carbon future (Turmes 2017, p. 36). That CCS was included served to dampen opposition from the coal industry although their primary interest had been excluded: clean-coal technologies that would include not only CCS but also more energy-efficient and less air-polluting generation technologies (EURACOAL 2006). The petroleum industry recognized that CCS could give a brighter low-carbon future for natural gas-based electricity, and storage of carbon in oil and gas reservoirs could represent new business opportunities.

Various renewable energy industry associations and companies teamed up to coordinate an agenda for technology development in Commission-established ETPs on solar PV, solar thermal power, wind-power, and biofuels. The nuclear energy industry, already with a privileged

[14] According to one involved staff member of DG Energy, hundreds of micro-tech companies lobbied for their solutions (Interview B).

R&I budget position in Brussels under the EURATOM Treaty, hoped for even higher budgets and needed to realize its plan of developing a new generation of fission reactors.[15] The first nuclear fission platform was proposed in 2005 and led to the final establishment in 2007 of the Sustainable Nuclear Energy Technology Platform (SNETP Undated).

When the Council and Parliament decided that energy efficiency should in fact be given focus in the Plan, the Commission in 2009 proposed the additional Smart Cities and Communities Initiative. This initiative did not primarily target industry as a beneficiary of R&I support but a wider group of actors, cities and municipalities. One group had great influence on this Commission decision, the Climate Alliance, formed in the early 1990s as a network organization established when a few German, Austrian and Swiss municipalities joined forces to protect the world's climate and rainforests. By 2000, 900 municipalities had joined and signed up to the Bolzano Declaration, committed to climate policy at the local level. The role of local-level action in fighting climate change was acknowledged by the EU when it adopted the Intelligent Energy for Europe (IEE) Programme in 2001, essentially a deployment support programme for renewable energy and efficiency solutions. Renewed acknowledgement came in 2008, when the Commission established the Covenant of Mayors as a new EU arena for local authorities to discuss contributions towards achieving the 2020 objectives, based on advocacy from, inter alia, the Climate Alliance (Climate Alliance 2018).

Summing up: industries mobilized ahead of the SET-Plan to secure their place in the future technology landscape of Europe. Key technology manufacturers and technology users strongly supported the idea of a SET-Plan, to get more of the budget allocated to industrial demonstration activities. There were also powerful voices arguing for a massive widening of priorities as compared to the selection made by the Commission. Reflecting their own R&I interests, industry associations differed in their preferences as to which technology areas should be given priority. And finally, manufacturing and electricity-supply industry actors succeeded in uploading their top priorities as EIIs. The ETPs emerged as the key link between the Commission proposal and industry mobilization, as indicated by the high correspondence between technology areas with such ETPs

---

[15] The distribution of R&I funds was a major issue—EWEA in its annual report pinpointed that nuclear research received 18 times more funding than windpower research, and conventional energy sources three times more (EWEA 2006, p. 14).

and the EIIs selected for the Plan. However, one priority that had enjoyed broad backing across various key industry actors did not make it to the Commission's EIIs proposal: to promote more efficient energy end-use technologies. The pressure from industries to include such technologies was weakened by strong disagreement on which specific end-use technologies should be given top priority among a host of alternatives.

4.2.3.2 The Research Community

Key research community actors with stakes in the SET-Plan were European universities and research institutions organized at the EU level in bodies such as the European University Association and the Commission's Advisory Group on Energy. The energy industry had DG Energy as its main interlocutor with the Commission; DG Research was main speaker for the research community. Implementation of the European Research Area, for which DG Research had main coordinating responsibility, would entail mobilizing European research institutions to foster greater mobility for researchers across the borders, coordinate research agendas and projects and link up with industry to co-develop research agendas that could result in new products brought to the market for selected technology areas, including energy.

The research community had first-hand knowledge of the research frontiers for various technology areas. The Advisory Group on Energy was important in assisting the Commission in inventorying possible technology areas for the SET-Plan and their level of maturity: to what extent the technologies would need further R&D efforts, were ready for full-scale industry-led demonstration, or primarily needed deployment measures to enter the market (Commission 2007d).

Key research community actors back in the early 2000s recommended a stronger strategic take on EU energy research, to remedy fragmentation and improve transparency and coordination between DG Research and DG Energy, responsible for managing Framework Programme budgets that targeted energy research and demonstration projects, respectively (Chap. 3). The research community had experienced how poor transparency and coordination within Commission services caused problems with project application (European Court of Auditors 1998). In 2005, a working group under the Advisory Group on Energy for FP6 reiterated the diagnosis of fragmented EU energy R&I and called for more rapid implementation of ERA objectives specifically for the energy sector—a new European Energy Research Area should be established. Recognizing the lack of strategic focus in EU energy R&I, the group recommended redirecting energy research towards visionary goals and market deployment of complete technologies—and that would call for a well-coordinated approach, as well as a pooling of the resources available at regional and national levels (ERAWOG 2005). A main conclusion was that the Commission should continue with its top-down actions to stimulate coordination in planning and funding of energy research, but should seek support of the member-states and all other stakeholders. The key message of the 2005 ERAWOG report was: 'Time has been lost. The real work of creating the European Energy Research Area must start' (p. 2). The Commission proposal for the SET-Plan followed up on these recommendations.

When the Commission tabled its draft of the SET-Plan in January 2007, another Advisory Group on Energy report, *Transition to a Sustainable Energy System for Europe: The R&D Perspective*, was used as reference for technologies that could be contemplated for the SET-Plan. This report documented 26 key energy-technology areas (16 electricity/ heat conversion and 10 transport sector areas) as future options. It also analysed end-use energy-efficiency technologies—but these were not presented in the communication, as the Commission concluded that the range was so extensive that a concise summary was impossible (Commission 2007c, pp. 11–12). This report noted various research-community interests linked to a very wide portfolio of technology areas, reflecting the engagement of the research community in research and development across this broad scope of technologies.

The Advisory Group on Energy continued to support the SET-Plan premise that a selection should be made. In 2006, it advised that the FP7 energy work programmes 'should be as highly concentrated as possible on a small set of crucial areas where real advances could have the greatest impact' on European competitiveness, security of supply or sustainability (Advisory Group on Energy 2006, p. 5).

While agreeing on such high-order principle, the diverse interests within the research community constrained the Advisory Group on Energy from recommending what specific technology areas should be selected as top priority for EU efforts. The 2006 report of the group referred to various splits internally in the group on what energy-*supply* technologies should be promoted and not promoted: 'Whilst most AGE members support the principle of focus on key priorities, there is less agreement on what the priorities should be, or perhaps more strictly, on which areas could be sacrificed to allow for greater concentration ... views differ about the relative priorities … Some concerns have been expressed about undue emphasis … to renewables at the expense of energy efficiency and conventional energy. Some are sceptical about the prospects for a hydrogen economy … Some see … Generation IV as the overriding priority for the fission programme. Other see renewables … as the top priority' (Advisory Group on Energy 2006, pp. 4–5).

The group managed to unite only on a very vague statement to the effect that energy efficiency and demand-side research should not have priority over the supply side, but not be neglected or downgraded in the favour of more 'interesting' supply-side research (Advisory Group on Energy 2006, p. 5). It recognized that delaying further progress in energy-efficiency research would be a serious mistake, although it was aware that there were already deployable energy-efficiency technologies available that could reduce energy consumption in Europe by as much as 30 per cent (Advisory Group on Energy 2006, p. 5).

While the 2005 Advisory Group on Energy ERAWOG-report recommended redirecting energy research in Europe towards market deployment, the 2006 report voiced considerable disagreement about the role and EU-funding of large-scale industrial demonstration projects: 'Some members of AGE see pilot demonstrations to be required to establish the credibility/provide the proof of principle of a new technology. On the other hand, they feel that large scale demonstrations often have the character of "more of the same" … As a consequence, some members of AGE advise that pilot demonstrations should be supported, but that large-scale demonstrations should not' (Advisory Group on Energy 2006, pp. 9–10).

Summing up, the research community played important roles in initiating the SET-Plan through advisory roles for the Commission and participation in the ETPs. First, the research community sided with industry in calling for the SET-Plan to raise the level of funds available for energy R&I in Europe and to improve coordination based on priorities. Secondly, the research community disagreed on what specific technology areas should be chosen for top priorities for energy R&I efforts. This reflected the fact that universities and research institutions had underlying interests in the very wide portfolio of technologies then at various stages of development in Europe. Thirdly, unlike industry, which supported broadly redirecting resources towards large-scale industrial demonstration projects, the research community held that resources were best spent on developing low-carbon technologies at lower stages of maturity. Finally, the proposed SET-Plan signalled a compromise between the interests of the energy industry and the research community, reflecting their different relationships with the Commission's DG Energy and DG Research. Resources should be allocated to projects at different stages of development; both groups would be given roles in steering implementation, through EIIs and EERA, respectively.

4.2.4   Summary

The Commission's initial selection of six priority technologies for the SET-Plan had focused on energy supply. In addition to member-state interests, this selection reflected existing cooperation with industry and the research community in European Technology Platforms. It also reflected the interests of well-organized incumbent energy and technology manufacturing industries as well as new renewable energy industry. The European Parliament and parts of industry and the research community pushed for wider priorities, including energy efficiency. They differed in their views on the relative roles to be given to large-scale demonstration of mature technologies in the short-run versus research and development of less-mature technologies for a longer time frame. This reflected an inherent tension in the SET-Plan that was not solved by a formalized compromise in the Council and Parliament decision on the Plan.

4.3   International Technology Markets

The SET-Plan was established in a context of competing international low-carbon technology investments and markets. As stressed in the Kok report on the revision of the Lisbon Strategy: 'Europe can gain a first mover advantage by focusing on resource efficient technologies that other countries will eventually need to adopt' (Kok 2004, p. 35). The report depicted the EU as facing a twin international competition challenge, from Asia and from the United States.

When the Kyoto Protocol was ratified in 2005, the Commission's DG Environment picked up the 'first mover' argument, linking it to climate change: 'As an example, the [European] countries that have taken the lead in promoting wind energy now have 95 per cent of the rapidly growing wind turbine industry. Looking forward, this kind of phenomenon could also emerge in other countries and other sectors … Competitive advantages will be enhanced if participation in a future international climate agreement is broadened and deepened' (Commission 2005b, p. 7). From then on, EU low-carbon technology-push and market-pull policies became directly linked to sustaining 'first-mover' advantages abroad. Strengthening competitiveness in international technology markets has been an integral element of the SET-Plan since its inception.

Before the SET-Plan was initiated, the United States and Japan were singled out as the main low-carbon technology competitors, followed by emerging economies like China, India and Brazil (Commission 2007d). Their market size, investment and research capacities were far exceeding those of most EU member-states. In 1977, the United States created a single department (ministry) for all energy-related matters in response to the first oil crisis. Through its system of ten national laboratories, the Department of Energy (DoE) funded more basic and applied scientific research than any other federal agency. In Japan, energy research enjoyed top priority, and was organized by the Council for Science and Technology Policy directly coordinated by the Prime Minister (JRC 2008).[16] The EU lagged both countries in total and energy R&I spending (public and private) relative to GDP (Commission 2007d).

The EU had a comparative advantage in publicly funded research on renewable energies. Compared to the United States, a larger share went to renewable energy, at the expense of fossil fuels. EU member-states also gave higher priority to renewables than Japan (JRC 2008). Moreover, the EU-25 were in the lead on renewable power deployment (solar PV, geothermal, biomass, wind, small hydro), (REN21 2007).

Bioenergy was highly prioritized by several EU countries. With Sweden and Finland at the front, the EU was among the leaders in biomass-based electricity generation. The United States, the largest investor in biofuels R&D for the transport sector (JRC 2008), became the world's leading fuel ethanol producer in 2006; Germany produced half of the world's biodiesel (REN21 2007). Solar PV stood out as the fastest growing power-generation technology in the world (REN21 2007). Japan had its world-leading solar PV industry organized around a few major corporations that were less dependent on public R&D than was the European PV

---

[16] Data in this section are based on JRC 2008. Due to limitations in Eurostat and IEA datasets, their numerical precision and comparability with the United States and Japan are surrounded by uncertainty. For example, only 17 out of 27 EU member-states were IEA members.

industry. The PV industry in Europe—led by Germany and the Netherlands—was characterized by small- and medium-sized companies that were dependent on public R&D subsidies (JRC 2008). In 2006, Germany accounted for half of the solar PV market, followed by Japan and the United States (REN21 2007).

The EU lagged the global leading position of the United States on High Temperature Solar Thermal Technologies; Germany, Spain and increasingly Italy were catching up. On the other hand, the United States and some emerging economies were catching up on windpower. Still, Europe accounted for 67 per cent of global installed wind capacity, with Germany in the lead (REN21 2007). Denmark, Germany and the Netherlands accounted for over 70 per cent of EU R&D in windpower (JRC 2008). Two-thirds of global windpower additions in 2006 were concentrated to five countries: the United States (2.5 GW), Germany (2.2 GW), India (1.8 GW), Spain (1.6 GW) and China (1.4 GW) (REN21 2007). Offshore windpower installations were emerging slowly, mainly in Europe.

Clean fossil-fuel technologies as well as CCS were becoming top priorities in the United States and in certain EU member-states (JRC 2008). The US DoE's carbon sequestration programme had increased from €8 million per year in 2000 to €50 million by 2006, whereas annual R&D budgets for the key EU players—Germany, United Kingdom, France—ranged between €10 and 20 million. Still, one of the world's leading aquifer CCS projects was located at Norway's offshore Sleipner field, and involved oil companies, the European Commission, IEA and DoE. Nuclear-fission R&D focused on safety, cost reduction operability and new reactor designs (JRC 2008). Japan was the clear R&D leader here, followed by the EU and the United States. Headed by France, which spent over 60 per cent of its overall R&D budget on nuclear, such research was among the first priorities for about one out of three member-states—corresponding to those with a nuclear-intensive power sector. The United States and Japan were more advanced when it came to standardization and energy codes.

Most developed countries also had market-pull instruments in place for renewables. After phasing out subsidies for solar power in 2005, Japan became the first market where solar became competitive without subsidies, due to the high price of electricity (JRC 2008). Thanks to subsidies, the wind and solar business accelerated in Europe, the United States and emerging economies. China, India and Brazil had begun implementing several renewable energy deployment policies, including feed-in tariffs (REN21 2007). In Brazil, policies established in the 1970s made ethanol competitive, by promoting fuel derived from home-grown sugarcane. However, spurred by the 2003 EU Biofuels Directive (2003/96/EC), the EU led the biodiesel market. The EU ETS would drive technology development for non-European companies that operated in Europe or sold their products there.

Thus, we find that the SET-Plan itself was partly motivated by global competition concerns and the prospects of new business opportunities. The EU had a drawback in its administrative fragmentation and non-aligned research strategies that reduced the capacity of its research base. The SET-Plan aimed to rectify precisely this situation and strengthen the EU's leadership in low-carbon technologies in the face of growing competition. Although such EU-external forces contributed to the initiation of the Plan, we find no strong indications that the selection of technology areas reflected relative competitive advantages. The Plan gave priority to technology areas where the EU already had gained a stronghold in international markets, but also areas led by other countries, as with solar PV and nuclear fission. The technology priorities of the Plan were motivated mainly by internal EU needs.

Since the adoption of the SET-Plan, substantial changes have occurred internationally in public R&D, low-carbon technology investments and installed capacity. The EU lost its former clear leadership on wind and solar to China (Chap. 6).

4.4   Conclusions

In Chap. 2, we outlined empirical expectations related to the Liberal Intergovernmentalism (LI) and Multilevel Governance (MLG) approaches as well as the contextual factors linked to international technology markets. Our main conclusion is that LI and MLG can explain different aspects of the making of the SET-Plan. In line with MLG, the Commission took the initiative and gave priority to technologies based on industry and research community networks already established. Industry and research interests excluded from the Plan then mobilized to get 'their' technologies included. The Plan formed part of an integrated package of climate and energy policies designed by the Commission to achieve targets and combine different interests within the Commission and among the EU member-states. Member-states and the European Parliament were under pressure to agree on the whole package. The European Parliament had been excluded from the SET-Plan and voiced opposition to its narrow focus and priority of large electricity producers.

In line with LI, the governance system for the SET-Plan and the mode for selecting technologies became decentralized, reflecting diverse member-state preferences based on national competence over energy R&I resources. The Commission had initially selected technologies that were largely in line with the R&I interests and priorities of the major member-states. Commission efforts to exclude technologies favoured by member-states triggered opposition and failed to succeed.

External factors played an important role in initiation of the SET-Plan. The Plan was motivated partly by concerns for global competition, and the prospects of new business opportunities in international markets. Such external concerns, however, were not what decided the technology priorities in the end. The initial selection of technologies mainly served to accomplish specific EU-internal targets and political needs.

The SET-Plan contained built-in and unresolved tensions. Diverse member-state preferences and lack of unity between the EU institutions and among non-state actors led to pressures for expanding the list of technologies selected. Further, DG Research, the research community and the European Parliament wanted greater focus on R&D in less mature and more long-term technologies, whereas DG Energy and the energy industry preferred a focus on mature technologies and large-scale demonstration that could contribute to achieving the 2020 targets. Such tensions became reactivated and affected implementation of the Plan—to which we turn in the next chapter.

References

Advisory Group on Energy. (2006, November 8). *Opinion on the 2007 Work Programmes Under Horizon 2020*. https://ec.europa.eu/research/fp7/pdf/old-advisory-groups/energy-wp-2007.pdf#pagemode=none. Accessed 19 June 2018.

Barnes, P. M. (2011). The Role of the Commission of the European Union: Creating External Coherence from Internal Diversity. In R. K. W. Wurzel & J. Connelly (Eds.), *The European Union as a Leader in International Climate Change Politics* (pp. 41–58). London: Routledge.

BusinessEurope. (2007, November). *BusinessEurope Comments on the Upcoming Draft EU Strategic Energy Technology (SET)-Plan*. Brussels: BusinessEurope.

Climate Alliance. (2018). *Covenant of Mayors on Climate & Energy.* http://www. climatealliance.org/activities/covenant-of-mayors.html. Accessed 20 June 2018.

Commission. (2000). *Towards a European Research Area.* COM (2000) 6, 18 January. Brussels: European Commission.

Commission. (2004, September 21). *Technology Platforms—From Definition to Implementation of a Common Research Agenda.* http://www.eirma.org/sites/ www.eirma.org/files/doc/documents/EU/EU-041014-TechPlat.pdf. Accessed 12 Sept 2018.

Commission. (2005a). *DG Joint Research Centre, Annual Report 2005.* Brussels: European Commission.

Commission. (2005b). *Winning the Battle Against Global Climate Change.* SEC (2005) 180, 9 February.

Commission. (2006). *Summary Report on the Analysis of the Debate on the Green Paper 'A European Strategy for Sustainable, Competitive and Secure Energy'.* SEC (2006) 1500. Brussels, 16 November.

Commission. (2007a). *Limiting Global Climate Change to 2 Degrees Celsius: The Way Ahead for 2020 and Beyond.* COM (2007) 2 Final, 10 January. Brussels: European Commission.

Commission. (2007b). *An Energy Policy for Europe.* COM (2007) 1 Final, 10 January. Brussels: European Commission.

Commission. (2007c). *Towards a European Strategic Energy Technology Plan.* COM (2006) 847 Final, 10 January. Brussels: European Commission.

Commission. (2007d). *SET-Plan Impact Assessment.* SEC (2007)1508/2. Brussels: European Commission.

Commission. (2012). *JRC Petten 50 Years: 1962–2012.* Luxembourg: Publications Office of the European Union.

Commission. (2016). *Innovative Financial Instruments for First-of-a-Kind, Commercial-Scale Demonstration Projects in the Field of Energy.* Report Written by ICF in Association with London Economics, September 2016. Brussels: European Commission.

Commission. (2017). *Highlights of the JRC, 50 Years in Science.* https://ec. europa.eu/jrc/sites/jrcsh/files/jrc_50_years_brochure_en.pdf. Accessed 19 June 2018.

Eikeland, P. O. (2011). The Third Internal Energy Market Package: New Power Relations Among Member States, EU Institutions and Non-state Sectors? *Journal of Common Market Studies, 49*(2), 243–263.

Eikeland, P. O. (2012). *EU Energy Policy Integration—Stakeholders, Institutions and Issue-Linking* (FNI Report 13/2012). Lysaker: Fridtjof Nansen Institute.

Eikeland, P. O. (2013). Electric Power Industry. In J. B. Skjærseth & P. O. Eikeland (Eds.), *Corporate Responses to EU Emissions Trading: Resistance, Innovation or Responsibility?* (pp. 45–98). Farnham: Ashgate.

EPPSA. (2007). *Annual Report 2006*, May 2007. http://www.eppsa.eu/tl_files/eppsa-files/3.%20Publications/Annual%20Reports/2007%20-%20EPPSA%20Annual%20Report%202006.pdf. Accessed 20 June 2018.

ERAWOG. (2005). *Towards the European Energy Research Area*. Recommendations by the ERA Working Group of the Advisory Group on Energy, EUR 21353. Brussels: Directorate-General for Research.

EURACOAL. (2006). *Contribution to the Fossil Fuels Forum Working Group on Coal as well as on the Green Paper: A European Strategy for Sustainable, Competitive, and Secure Energy*. https://www.slideshare.net/EU-2020/eura-coal. Accessed 13 Sept 2018.

EurActiv. (2005a, July 20). 'High Noon' for EU's Environment Policies. *EurActiv*.

EurActiv. (2005b, December 21). Crunch Time for EU Environmental Policies. *EurActiv*.

Eurelectric. (2007, July). Eurelectric Views on the EU Strategic Energy Technology Plan. *Eurelectric*.

Eurelectric. (2008, April). Eurelectric Comments on EU 'Strategic Energy Technology Plan'. *Eurelectric*.

European Court of Auditors. (1998). *Special Report No 17/98*, on Support for Renewable Energy Sources in the Shared-Cost Actions of the JOULE-THERMIE Programme and the Pilot Actions of the Altener Programme Together with the Commission's Replies. *OJ*, C356/39, 20 November.

European Parliament. (2005). *Tony Blair Tells MEPS How Europe Should Face Up to Globalisation*. Info Press Service, Directorate for the Media. http://www.europarl.europa.eu/sides/getDoc.do?pubRef=-//EP//NONSGML+IM-PRESS+20051103IPR02004+0+DOC+PDF+V0//EN&language=BG. Accessed 25 June 2018.

European Parliament. (2008a). *Draft Report on the European Strategic Energy Technology Plan*. Committee on Industry, Research and Energy. Brussels, 17 March.

European Parliament. (2008b). *Amendments to the European Strategic Energy Technology Plan*. Committee on Industry, Research and Energy. Brussels, 8 May.

European Parliament. (2008c). *Opinion of the Committee on the Environment, Public Health and Food Safety for the Committee on Industry, Research and Energy on the European Strategic Energy Technology Plan*. Brussels, 3 June.

European Parliament. (2008d). *European Parliament Resolution of 9 July 2008 on the European Strategic Energy Technology Plan* (2008/2005(INI)). Strasbourg: European Parliament.

European Parliament. (2014, October). *The Open Method of Coordination*. Note from European Parliamentary Research Service. http://www.europarl.europa.eu/EPRS/EPRS-AaG-542142-Open-Method-of-Coordination-FINAL.pdf. Accessed 19 June 2018.

EWEA. (2006). *Powering Change, Annual Report 2006.* http://www.ewea.org/fileadmin/files/library/publications/reports/EWEA_Annual_Report2006. pdf. Accessed 20 June 2018.

JRC. (2008). *Energy Research Capacities in EU Member States* (JRC Scientific and Technical Reports, EUR 23435 EN – 2008). Petten: Joint Research Centre of the European Commission.

Kok, W. (2004, November). *Facing the Challenge: The Lisbon Strategy for Growth and Employment* (Report from the High Level Group Chaired by Wim Kok). Luxemburg.

Orgalime. (2006, September 22). *Orgalime Position Paper on Green Paper on a European Strategy for Sustainable, Competitive and Secure Energy.* COM (2006) 105 Final, Brussels. www.orgalime.org/sites/default/files/position-papers/Orgalime%20position%20GP%20Energy_Sep06.pdf. Accessed 03 Oct 2018.

REN21. (2007). *Renewables 2007 Global Status Report.* Renewable Energy Policy Network for the 21st Century. Paris: REN21 Secretariat.

SETIS Magazine. (2017, November). *Looking Back at 10 Years of Forward Thinking, the SET-Plan—We Need a Place Now More than Ever for Concrete Collaboration Between Countries.* https://setis.ec.europa.eu/setis-reports/setis-magazine/looking-back-10-years-of-forward-thinking-set-plan/we-need-place-now. Accessed 14 June 2018.

Skjærseth, J. B., Eikeland, P. O., Gulbrandsen, L. H., & Jevnaker, T. (2016). *Linking EU Climate and Energy Policies: Policymaking, Implementation and Reform.* Cheltenham: Edward Elgar.

SNETP. (Undated). *SNETP Annual Report September 2007–December 2008.* http://www.snetp.eu/wp-content/uploads/2014/02/SNETP_Activity_Report_2007-2008.pdf. Accessed 13 Sept 2018.

Turmes, C. (2017). *Energy Transformation: An Opportunity for Europe.* London: Biteback Publishing.

Wejnen, M., Greer, H., Forlani, F., Wagner, U., & Salminen, P. (2000). *Five-Year Assessment Report Related to the Specific Programme: Energy, Environment and Sustainable Development Covering the Period 1995–1999.* Brussels: Commission. https://ec.europa.eu/research/evaluations/pdf/archive/five_year_assessments/five-year_assessment_1995-1999/fp5_panels_final_report_energy_environment_and_sustainable_development_2000.pdf#view=fit&pagemode=none. Accessed 19 June 2018.

CHAPTER 5

Explaining Implementation of the SET-Plan

Why did the EU largely fail in its attempts at strengthening, focusing and giving coherence to low-carbon energy research and innovation, as prescribed by the SET-Plan? Here we seek explanations in the role of member-states, EU institutions and non-state actors, with EU market-pull policies as contextual factors.

5.1 ROLE OF MEMBER-STATES

Implementation of the SET-Plan would require high commitment by the member-states and their effective execution of assigned tasks. Such commitment would be even more important since the SET-Plan was not legally binding. The implementation challenges documented in Chap. 3 can partly be explained by lack of member-state commitment and involvement in SET-Plan implementation arenas.

The Steering Group was the most important arena for the member-states. The Terms of Reference for the Steering Group were agreed in July 2008 (Commission 2008)[1]: the group was to work to reinforce coherence

---

[1] The mission statement of the Steering Group mirrored the objectives of the plan: to contribute to and foster the implementation of a collaborative European Energy Technology Policy to accelerate the development and market uptake of clean, sustainable and efficient energy technologies, as a fundamental pillar of the EU's response to the interrelated challenges of climate change, security of energy supply and competitiveness (Commission 2008, p. 1).

© The Author(s) 2020
P. O. Eikeland, J. B. Skjærseth, *The Politics of Low-Carbon Innovation*, https://doi.org/10.1007/978-3-030-17913-7_5

93

between national, European and international efforts by steering the implementation of European technology policy. Further, it should promote joint actions and optimize R&I efforts, including coordination between the EIIs and EERA, and provision of data for the monitoring and review system SETIS.

The Steering Group was composed of representatives from the Commission, the 28 EU member-states and Iceland, Norway, Switzerland and Turkey. The Commission was to chair the Steering Group, serve as secretariat and manage SETIS. Running expenses would not be covered by the Commission—underscoring the central role of the member-states in governing the Plan. Each member-state was to designate two high-level representatives from the energy and research authorities having 'sufficient authority and knowledge to take positions on Community and national research and innovation investments' (Commission 2008, p. 3). The intention was to create a close link between the Steering Group, national and EU funding sources aligned to the priorities of the SET-Plan.

The decision-making procedure was based on consensus, regionalization and voluntary participation. Groups of member-states could voluntarily agree specific joint actions in line with their national situation and R&I priorities. The Steering Group could establish working groups to examine issues and invite other interested parties from industry, the research community or the financial sector on an ad hoc basis. Meetings of the Steering Group would normally be prepared by 'sherpas', designated by the member-states to meet on a more permanent basis and cooperate closely with the Commission Secretariat. In practice, then, the Steering Group was set up as an arena for member-states to explore and exploit shared low-carbon R&I interests.

Minutes from early Steering Group meetings show that less than half of the member-states were active in giving presentations. These meetings appear to have dealt mostly with procedural matters, and outcomes were categorized as 'Actions' and 'Decisions'. One 'action' resulting from the Steering Group meeting on 27 October 2011 was that the Steering Group decided to have systematically EERA on its agenda and that EERA should look for interactions with Joint Programmes and the EIIs (Commission 2011). This indicates that the three years which had passed since the Plan was decided had produced little systematic implementation.

Also indicative were the reflections that emerged in discussions of the Commission's 2009 communication for funding (Council 2010). The Council, which had restricted the Plan's authority, appeared impatient to accelerate implementation, stressing that the EIIs should move from the Technology Roadmaps to the operational stage, and that the activities of EERA should be made consistent with those of the EIIs. Further, it requested the Commission to develop SETIS to provide a robust technology-neutral planning tool, reflecting the state of the art of individual technologies and their anticipated technological development and market potential. Concerning the Steering Group, the Council requested the member-states and the Commission to 'streamline the existing institutional steps and to give due regard to the central role of the SET Plan Steering Group...for the conception, launch and implementation of the various SET Plan activities, in particular for the EIIs and the research programmes of the EERA.' (Council 2010, p. 5).

Some member-states presented their own low-carbon R&I projects and invited other member-states to join. At the July meeting in 2012, for example, Germany informed about a smart grid project and a joint Finnish–German call (energy efficiency, smart grid, energy storage, batteries for stationary and mobile applications, fuel cells and hydrogen). The United Kingdom invited others to join in a wind and storage project, and Austria on a Smart City project (Commission 2012a). Member-states also informed about their various national approaches to improve R&I cooperation: the Netherlands noted the close link between the SET-Plan and the Dutch Energy Top Sector approach, and Germany spoke of the 'Berlin Model'.[2] While these examples illustrate some of the intended functions of the Steering Group, they also show that, despite the ambition for the group to steer low-carbon technology priorities strategically, it was in fact becoming a forum for information exchange among a few member-states.

Commission representatives tended to do the talking: most national representatives were entirely silent (Interview D). Many member-state governments were not interested in coordinating their national R&I resources for joint actions on the SET-Plan priorities, although a few representatives, from countries with abundant energy R&I resources, were committed to joint actions, as was illustrated by the few and poorly resourced ERA-NET initiatives for the period 2007–2013 (Interviews D, F).

---

[2] Based on three steps: (1) potential project partners from several member-states identify a joint research project and present a draft to their national funding agencies; (2) upon positive evaluation by all national funding agencies, project partners submit a full proposal; (3) after decision by the national funding agencies, project partners request the Commission for additional support.

However, the process of negotiating the joint actions was cumbersome and resulted in very little money (Interviews D, G).[3] One important problem was that many member-states had no programmes for supporting demonstration of mature technologies (Interview F). Even the most active representatives might be unable to pledge substantial joint actions to support the EIIs at the EU level. Some member-states added constraints that national funds committed to joint actions under ERA-NET should be distributed exclusively to national industries and research communities (Interviews F, G).

There was another grouping, of countries with abundant energy R&I resources that did not commit to joint action; and a third major group representing most of the energy R&I resource-poor CEECs (Interview D). This third group, which tended to keep silent, included countries that were interested in only a limited selection of SET-Plan priorities. Some, like Poland, prioritized R&I in energy technologies that were not part of the Plan, such as clean coal (Skjærseth 2018).

Dissatisfaction with how the Steering Group was working surfaced at the group meeting in September 2012 (Commission 2012b). The most committed representatives raised questions about the cooperating procedures or 'working modalities', stressing the need to enhance the role of the group. It was also held that the 'sherpa' meetings that prepared the group meetings failed to devote enough time for political agreement at the national level. Moreover, national representatives varied significantly in authority to pledge resources, making the link between the Steering Group and national R&I funding weak. The Commission reiterated in 2015 the need for a new type of SET-Plan management (Commission 2015). This shows that little had happened, and the Steering Group was about to run out of steam. The Commission's JRC also stopped publishing minutes of Steering Group meetings at the SETIS site.

The SET-Plan member-states had wider responsibilities, however. A next important task was to engage in EII teams to come up with operational plans for the Technology Roadmaps and attach national funding to realizing these. The EIIs represented the sharp end of the SET-Plan tasked with developing industrial-driven research and innovation for large-scale demonstration projects (see Table 3.1, Chap. 3). Under the guidance of the Steering Group, the EIIs were led by 'EII Teams' composed of, and

---

[3]According to DG Energy coordinator of the Wind Energy EII, the member-states debated an ERA-NET for two years before agreeing on very little funding (Interview G).

appointed by, member-states, industry and the Commission. The teams were to provide a platform for planning actions to implement the initiatives. This included defining goals, developing actions, identifying investment needs, putting activities into operation and monitoring them. The teams were also to address cross-cutting issues and synergies to other EII teams.

The EII teams apparently did not work very well. They produced implementation plans, but struggled to put these into operation. A major problem was low member-state participation and commitment. Key member-states with relevant R&I resources were not engaged in the EIIs on wind, solar, bioenergy, sustainable nuclear and CCS (Commission 2013a). The member-states did not show clear commitment to strategic planning and joint action for investments (Commission 2013a, p. 8). Several national governments were represented by persons not employed by government or national funding agencies (Interview O). Given the lack of pledged national resources committed, the teams had to devote considerable effort to agreeing and setting priorities on the basis of more limited EU-level resources like FP 7, NER 300 and the European Energy Recovery Programme (Commission 2013a). The SET-Plan had only limited success in aligning its priorities to these funding sources (see Chap. 3).

The respective roles of the member-states and the Commission in the EII decision-making process and between the Steering Group and the EII teams were unclear. This also weakened the links between the EIIs.[4] The EIIs rarely triggered additional investments, and then only marginally (Interview D).

The member-states were also tasked with steering and supporting the research community to set up joint programmes under EERA and align these with the EIIs. Whereas the EERA attracted wide participation from European universities and research institutions and contributed to the consolidation of national research capacities, the links with most EIIs remained weak, and any joint programmes were described as largely 'virtual' (Commission 2013a). Evidently, national public research-funding bodies had not delivered and attached funding to the joint programmes, leading the Commission to conclude 'EERA and the ... EII are not delivering to the level required to move the SET Plan forward' (Commission 2015, p. 5).

---

[4] For example, the EII on electricity grids and other EIIs on renewable power production depending on grid innovation and deployment were not coordinated (Commission 2013a).

A final important task for the member-states was to provide R&I information to SETIS, essential for governing the SET-Plan. Efforts to monitor member-states' programmes, investments and initiatives in support of energy research and innovation were not very successful (Commission 2013a). A key challenge was insufficient reporting on national research and innovation priorities and investments (Commission 2015, p. 4). For example, the capacity mapping of R&I investment in SET-Plan technologies lacked data for several member-states for various indicators, such as trends in investments (JRC 2015). Data came largely from the IEA, but not all EU countries are members.

Most member-states gradually became more engaged in energy research and innovation linked to the SET-Plan priorities (JRC 2015). In 2015, resources were still concentrated in the North, but countries like the Czech Republic, Poland and Slovakia increased their low-carbon R&I funding. Thus, although the potential for pooling resources through the SET-Plan has grown, it has largely remained unexploited. This has also been underlined by the Commission: 'although MS do share common industrial and research interests, their commitment to the SET Plan is suboptimal' (Commission 2015, p. 4).

In summary, most member-states lacked commitment to the Plan and many did not actively participate in the Steering Group, the EIIs, EERA and SETIS. This made it difficult for the member-states to join forces to strengthen, focus and give coherence to low-carbon energy research in Europe.

5.2   Role of EU Institutions and Non-State Actors

5.2.1   The Commission

As Secretariat for the SET-Plan, the Commission had important functions, like mustering commitment and pushing implementation. By 2009, it had established the Steering Group, the EIIs and the EERA. With no possibilities of enforcing cooperative behaviour on these new arenas, the Commission had to rely on 'soft' governance measures to ensure commitment. It could provide EU-level co-funding incentives, within the limits set by EU programmes. It could seek to hold member-states accountable to the Plan by benchmarking their performance through information collected by SETIS. And it could seek to muster commitment by making the

Plan visible on the political agenda. An important instrument has been the convening of annual SET-Plan conferences organized by the rotating EU Council Presidencies.[5] But these conferences have not attracted much attention to the SET-Plan from the public, organizations or the media.[6]

Commission staff took an active role as chair of the Steering Group aimed at mustering member-state government representatives to commit for joint actions (Interviews D and F). To create stronger incentives for the member-states, the Commission adjusted the ERA-NET mechanism so that it would allow EU programmes to co-fund and take part in burden-sharing. The number of joint actions agreed in the energy area increased, but the little funding pledged by the member-states proved insufficient to attract industry investments. The Commission also took a lead role on industry-chaired EII teams. Commission staff drafted the Technology Roadmaps that outlined research and demonstration tasks and resource needs towards 2020 and decided on the number and type of actors to give input (Interview Q). However, the Commission failed in getting private and public co-funding pledged for the EIIs.

The Commission further facilitated the establishment of the research pillar of the SET-Plan: the EERA. Co-initiators were ten leading public research institutes, supported by the European University Association (EUA) and the Heads of European Research Councils (EuroHORCS)— both research-executing and research-funding agents. Membership quickly mushroomed and today includes nearly all European universities: the latest updated list shows some 200 universities/research institutions (EERA 2018). A central task for EERA has been to coordinate research efforts in joint transnational programmes aligned with SET-Plan priorities. However, the Commission exercised little control in steering this area or the process of establishing joint programmes (see below).

---

[5] These conferences are opened by high-level host-country politicians; keynote speakers have included the energy and research Commissioners; and several hundred stakeholders and representatives from EU and national institutions across Europe have attended. Nine conferences were convened between 2008 and 2017: Paris (2008), Stockholm (2009), Madrid (2010), Brussels (2010), Warsaw (2011), Dublin (2013), Rome (2014), Luxemburg (2015) and Bratislava (2016).

[6] A search on ENDS Europe Daily, which follows EU climate and energy policies closely, brought up approx. 30 articles mentioning the SET-Plan. In comparison, since 2007, the Renewable Energy Directive and the Emission Trading System have featured in some 240 and 400 articles, respectively. This may give a rough indication of relative saliency.

The Commission's JRC had an important role in ensuring that SETIS was able to monitor and review implementation of the SET-Plan. As noted, the JRC faced major challenges regarding the lack of commitment on the part of member-states and industry to report R&I data. That the Commission lacked updated and reliable data on resource allocation to the SET-Plan impeded the detection of resource gaps for the various EIIs.[7]

On the other hand, the Commission's role in implementation was far broader than merely chairing coordination efforts by member-states, industry and the research community. Of fundamental importance was preparing new EU-level funding mechanisms and managing those already established, to ensure alignment with the priorities of the SET-Plan. As to new funding programmes, the Commission had the prerogative of drafting, whereas the European Parliament and Council would decide on the programmes, budgets for energy R&I and broad lines of allocation to technologies and project types. Within the limits set by these decisions, however, the Commission would have some leeway to coordinate and pool resources managed by the services and sub-services under various EU programmes and sub-programmes.[8]

The core EU R&I funding programme, FP7, had already been established before the SET-Plan, with broader priorities than those of the SET-Plan. Unsynchronized EU budget cycles proved problematic for the Commission's implementation efforts. The Commission was also dependent on the European Parliament, which had resisted reallocation of FP7 resources to SET-Plan priorities.

Fresh funding opportunities came with the EEPR and NER 300, but here the Commission had limited success in influencing these in line with the SET-Plan's priorities (see Chap. 3). Concerning EEPR, the Commission's initial proposal from 2008 was poorly linked to the SET-Plan, indicating coordination problems internally in the Commission.

---

[7] Lack of data is repeatedly reported in all JRC capacity maps; see also Commission (2013a).

[8] The establishment of the 'Smart Cities' initiative illustrates the Commission's limitations in overseeing implementation. The Environment Council in 2013 decided that the Commission could not take funding decisions in areas where the Parliament and the member-states had competences, such as the FPs. This built on previous Council and Parliament signals that the SET-Plan was not to have financial implications for existing EU funding programmes: the initiative was to be restricted to a platform for monitoring of actions, not for receiving financial contributions from R&D funds (Council 2013).

In 2009, the European Council decided on a closer link (to fund SET-Plan demonstration of CCS and windpower) and the European Parliament ensured that the fund would also fit with the Smart Cities and Communities Initiative. Hence, the priorities, procedures and eligibility criteria were decided by the Council in cooperation with the European Parliament.

NER 300 was based on revenues from auctioning of EU ETS emission allowances. It was administered by DG Environment/DG Climate Action, applied by this DG to strengthen its own capacity to manage low-carbon energy technology demonstration funds, a task that had otherwise rested mainly with DG Energy (Interview S). The two DGs had a history of rivalry, but relations had improved (Chap. 4). However, NER 300 was not well coordinated with the SET-Plan. Monitoring and review were not integrated in SETIS; and eligibility criteria made the fund open to more technologies and different project types than those prioritized by the Plan. The size of the fund contracted seriously after the collapse of the ETS allowance prices from 2009 (see below). Importantly, NER 300 settled that co-funding with EEPR was allowed, but the funds awarded would be subtracted from awards under NER 300. This constrained the Commission from creating a higher 'critical mass' of EU public resources to leverage funding from industry. The Commission had to rely on co-funding from the member-states, which failed for many projects.

All this indicates differing interests and lack of coordination internally in the Commission, as well as lack of decision-making authority that impinged on the Commission's opportunities for implementing the SET-Plan. Varying eligibility criteria for different project types and restrictions on coordinated funding made alignment difficult, even for programmes that overlapped thematically.

FP8, Horizon 2020 (2014–2020), brought new opportunities to allocate more funds in line with SET-Plan priorities. The Commission proposed an energy budget that was upscaled compared to FP7 and with specific earmarking for SET-Plan implementation for the Risk-Sharing Finance Facility aimed at large-scale demonstration projects. However, diverging interests within the Commission, temporarily put at rest when the SET-Plan was made, now weakened its ability to align SET-Plan priorities with FP8. The built-in tension in the SET-Plan between short-term priority to mature technologies and long-term priority to less-mature technologies became evident in the FP deliberations.

DG Research had responsibility for drafting Horizon 2020, and it catered to much wider research interests than DG Energy. DG Energy had primary responsibility for the SET-Plan, particularly the EIIs and their contribution to realizing short-term energy-policy goals for 2020. The research community was mainly occupied with pre-competitive research and a broad portfolio of technologies at varying levels of maturity. Industry had diverse interests related to size and preferred technologies, but a common interest in allocation to industrial-led projects. Horizon 2020 struck a balance, with its three major budget posts—*excellent science, industrial leadership and societal challenges*—and with specific arrangements for promoting the participation of small- and medium-sized enterprises (SMEs).

While Horizon 2020 would cater to a broader portfolio of technologies than the SET-Plan, the Commission had additional opportunities for alignment in the drafting of work programmes and calls for projects, within the limits set by the Council and European Parliament when deciding the programme. The Commission initiated internal work to improve coordination and coherence between energy R&I funds management by DG Energy and DG Research and other services (Commission 2017a, p. 794).[9]

The division of responsibilities between DG Research and DG Energy changed. Now both services would assume responsibility for funding the full R&I chain: DG Energy as regards the energy demand side—especially projects on energy efficiency, energy system (grids and storage) and Smart Cities and Communities; and DG Research as regards the energy supply side—especially renewable energy and decarbonization of fossil fuels (Interview D; Commission 2017a, p. 744).[10] Wider interservice groups were established with representatives from all relevant parts of Horizon 2020, to discuss the strategic approach of the work programmes. Despite these efforts, incoherence remained within and between EU programmes, stemming from different interests and absence of a central monitoring system within the Commission to ensure information flows among the services (Commission 2017a, p. 798).

---

[9] DG Research introduced a new project classification system under Horizon 2020 based on Technology Readiness Levels (TRLs) to specify the maturity level of technologies for projects to be called for under the work programmes (Interview B). This brought greater transparency and thereby opportunities for ensuring that work programmes and projects were more coherent with the SET-Plan.

[10] As the principle of equal budget split between the two DGs remained in place, the flexibility of the programme to focus more resources on either the supply or the demand side also remained limited (Commission 2017a, p. 744).

In summary, the Commission aimed at taking a leading role in implementing the SET-Plan. However, it proved unable to execute this role effectively through any of the Plan's new arenas and mechanisms. Main reasons were lack of decision-making authority, the difference in eligibility criteria between various EU funding programmes, unsynchronized budget cycles and diverging R&I interests within the Commission. Lack of internal unity made the Commission vulnerable to conflicting demands from others and weakened its impact as regards aligning the SET-Plan with EU funding programmes.

5.2.2   The European Parliament

The European Parliament had no formal role in the SET-Plan governance system, but would affect implementation by its role as co-decider of EU funding programmes. It thus had authority to affect actual energy R&I priorities, to what extent EU-level funds should be directed towards SET-Plan technologies, and towards the realization of large-scale demonstration projects.

The EP's response to the Commission's 2009 communication on funding the SET-Plan illustrates why alignment between EU funding programmes and the SET-Plan became difficult. The EP requested new criteria for selecting priorities for the Plan and announced future EU funding that could go counter to SET-Plan priorities.

The first response of the EP came in the form of two ITRE Committee reports. The Committee was split on one point: Left and Green MEPs did not agree that nuclear energy should be viewed as 'sustainable' and proposed renaming the 'Sustainable Nuclear Energy EII' to 'Nuclear Energy EII' (European Parliament 2010a, b). The Committee reaffirmed that the six EIIs were all welcome, but added that some of them should be advanced faster than others. Because of the economic crisis, priority should go to technologies with greatest potential for job creation, like the EIIs on solar power, windpower and biofuels (European Parliament 2010a, b). The Commission had estimated that between 200,000 and 250,000 new jobs could be created from each of these initiatives (Commission 2009a).

The ITRE Committee further urged for SET-Plan funding programmes to extend opportunities for various groups of actors. Technology manufacturers should be allowed to apply directly for SET-Plan funding under EEPR and NER 300, not just as members of consortia with energy utilities; and SMEs should get better access to public grants and loans in the

EU. This shows that the ITRE Committee was not content with how the SET-Plan sought to steer funding towards large-scale industrial demonstration projects to be conducted by major incumbent energy utilities (Chap. 3). It wanted redistribution of opportunities towards SMEs, job creation and energy consumers (European Parliament 2010a, b).

The plenary debate on 11 March 2011 added conflicts over the SET-Plan. Green Party group representative Claude Turmes set the tone, claiming that the SET-Plan was out of tune with actual investments in wind, solar and biomass by earmarking greatest share of EU funding to CCS and nuclear (European Parliament 2010c). Other MEPs, particularly from the CEECs, strongly supported the inclusion of nuclear power, but also questioned the SET-Plan's overarching focus on low-carbon technologies.

Nevertheless, the funding plan communication was adopted in the Parliament, by 444 votes to 88, with 32 abstentions. Although the proposed renaming of the nuclear power EII was not adopted, it was agreed that the SET-Plan should give more priority to technologies that provided the most opportunities for job creation and should ensure more support to SMEs and local-level actors (European Parliament 2010d). The debate and voting in the Parliament made evident that many MEPs remained critical to the SET-Plan and continued to request changes in what technologies and actors the Plan should focus on.

Later EP discussions and decisions on EU energy-funding programmes show that the conflict lines persisted, with real effects on alignment between the SET-Plan and EU-level programmes. The ITRE Committee report of 20 December 2012 contained hundreds of proposed amendments to Horizon 2020 (European Parliament 2012). One called for earmarking 75 per cent of the energy budget to R&I on renewable energy, energy efficiency, smart grids and energy storage. Another proposal called for funding a broader range of technologies than those prioritized by the SET-Plan. It was also suggested that the FP budget should not be applied at all to support the SET-Plan.

The Parliament's final resolution successfully called for clearer priority to non-fossil fuel projects under the Horizon 2020 Program to fund a wider range of low-carbon energy projects than initially selected for the SET-Plan. The Parliament also succeeded in earmarking funding for SMEs (thereby smaller-scale projects) and less-mature technologies, rather than concentrating resources on the demonstration of a few more-mature low-carbon technologies (European Parliament 2013).

The Parliament made similar changes to the EEPR. It effectively called for including smaller-scale energy efficiency and a larger portfolio of renewable energy projects in addition to CCS and offshore wind demonstration projects, arguing that energy efficiency and renewables would create most jobs by SMEs at local level. MEPs also managed to get the NER 300 expanded to fund small renewable-energy demonstration projects, beyond the priorities of the SET-Plan (Åhman et al. 2018).

The reform of the SET-Plan was in line with much of what the EP had wanted from the start: an extension of the scope of technologies to be given priority for EU R&I, including energy-system and energy-efficiency technologies that would benefit more sectors, and a greater number of renewable energy technologies and technologies that would promote innovation and job creation at the local level.

To sum up, the European Parliament contributed to strengthening EU low-carbon energy R&I, by pushing for higher EU-level budgets, but it weakened the opportunities for consolidating the SET-Plan around its initial selected technologies and large-scale demonstration projects. MEPs disagreed on what technology areas, project types and actors should be given priority in the Plan and EU funding programmes. While many MEPs supported more funding to SET-Plan industrial demonstration projects, others wanted more EU support to go to smaller SME projects and less-mature technologies with longer-term perspectives. The result was a political compromise, where the initial priorities and criteria of the SET-Plan were retained, and new ones were added.

5.2.3 Industry and Research Community

European industry affected SET-Plan implementation and contributed to the problems noted. Industrial companies were assigned a key role in realizing large-scale demonstration projects under the EIIs. Selected industry partners gave input to the Technology Roadmaps and agreed on the implementation plans (see Table 3.1, Chap. 3). However, although industry generally supported the Plan, several industry associations disagreed as to its top technological priorities (Chap. 4). This contributed to the lack of commitment to prioritize and make final investment decisions on large-scale demonstration projects, which affected some EII technology areas more than others (Chap. 3).

The financial crisis that unfolded from autumn 2008 made industry more averse to risky investments, and a fall in industry energy R&I funding resulted (Commission 2009b). Pressures from member-states, the EP and industries for developing technologies not included in the SET-Plan contributed to diverting funds away from the prioritized technologies and large-scale demonstration projects (Commission 2016). This combination of risk aversion and scattered funding weakened implementation of the EIIs.

Most EII teams did not function well, but for various reasons (Commission 2013a). For some EIIs (like wind and solar), industry was represented by European associations—not companies with resources for investments in the SET-Plan technologies.[11] The nuclear-power EII had weak funding commitment from industry and member-states. No near-term impact was expected on finalizing demonstration plants for Generation IV reactors by 2020, as had been promised in the Technology Roadmaps (Chap. 3). The bioenergy initiative attracted strong direct involvement of industry and the existing technology platform—but with more focus on demonstrating biomass-based heat and power, and less on advanced biofuels, as initially planned. Concerning the European Electricity Grids Initiative, headed by the European transmission system organization, ENTSO-E, there was a lack of connection with the Smart Grid European Technology Platform and communication technology industry. This was an industry vital in assisting development of Smart Grid solutions. Further, the CCS EII had a very broad focus on different CCS technologies, which indicated problems in agreeing on priorities for implementation. Thus, the EIIs failed to evolve as a central arena for industrial companies to commit own funding and mobilize external funding.

The diverse research community affected implementation in various SET-Plan arenas. The research community oversaw the EERA, which had been set up to propose joint programmes to be funded directly by national public bodies (research councils, ministries and others) outside the EU's main funding programmes.[12] EERA reported the first joint programme in 2010 (to date, 17 programmes have been adopted). The programmes included technology areas not initially prioritized by the SET-Plan. EERA

---

[11] Some EII teams included companies, but persons representing these came from the communications department and not the investment department (Interview O).

[12] An EERA Executive Committee, consisting of a minor group of research institutions, would first approve joint programmes. The Commission, EUA and EuroHORCS would have only observer roles (Research Council of Norway 2011).

had many participants (European universities and research institutions) with diverse interests in promoting a far wider portfolio of technologies. Many of the joint programmes were as noted largely 'virtual', focused on publicity and positioning rather than real coordination of research personnel and resources. The research community had weak links with the EII teams, and the joint programmes were poorly aligned with the implementation plans of the EIIs (Interview D).

The research community influenced implementation also by lobbying decision makers on EU R&I funding programmes. The Advisory Group on Energy, a key research community actor, first gave inputs to Commission working programmes set up under FP7 and disagreed with the heavy focus on large-scale industrial demonstration projects: 'Rather than seeking to re-direct the programme's annual call towards large industry, AGE considers that it could be a better strategy to modify the calls to encourage demonstration projects from SMEs and other smaller players, leaving large industries to collaborate with FP support through the "new instruments" such as JTIs and European Industrial Initiatives' (Advisory Group on Energy 2008, pp. 6–7).

The research community gave input to the Commission's drafting of Horizon 2020, and recommended channelling more resources to basic and applied R&D instead of industrial demonstration projects—allegedly because the scope for cost reductions would be highest for projects in the early stages of development (EERA, October 2010). This echoed tensions inherent in the SET-Plan as regards the relative distribution of funds for EERA joint basic research at universities and support to industrial demonstration of more-mature technologies within the EIIs. Decision makers—the Parliament in particular—lent an ear to the research community.

Industry and the research community also influenced the SET-Plan reform. Responses to the reform consultations show support from both groups for redirecting the focus towards energy-systems and energy-efficiency solutions (Commission 2013b). This reflected a redirection of R&I priorities already underway, aimed at responding to the new energy-system challenges resulting from massive investments in renewable electricity in many member-states. The Commission also asked stakeholders whether the initial six SET-Plan technologies should be retained as a focus for the revised Plan; responses differed widely across technology areas. Many respondents from industry and the research community saw continued R&I in solar power as important (26 per cent), whereas only 13 per cent held that CCS should continue as a priority (Commission 2013b, p. 14). And indeed, the SET-Plan reform came to accord higher priority to solar power and lower priority to CCS.

Clear differences were recorded between the research and industry communities as to preferences for the type of projects to be given priority in the revised SET-Plan. Industry favoured large demonstration projects; the research community saw basic and applied research as the most essential.[13] The research community appears to have prevailed with assistance from DG Research and the European Parliament: large-scale demonstration became less important in the revised Plan.

5.2.4   Summary

The roles and activities of EU institutions and non-state actors can help to explain the challenges entailed in strengthening, focusing and giving coherence to low-carbon energy research through the SET-Plan. The Commission played an active role in implementation, but lacked the authority to decide on funding from different programmes to realize the Plan's objectives. This resulted in a mismatch between SET-Plan priorities and which technologies and project types were eligible for support from various EU-level programmes. Moreover, budget cycles were not synchronized with the SET-Plan, and diverging interests within the Commission weakened its influence. The European Parliament had co-funding authority, but it was not part of the SET-Plan and had different priorities. MEPs successfully pressed for priorities that were wider than those of the SET-Plan. Weak, but varying EII realization can partly be traced back to the low commitment on the part of industry, due to scattered public funding, risk-averseness aggravated by the financial crisis, and low-level participation. The academic and industry communities exhibited little, but varying cooperation in joint strategic planning of programmes and projects.

---

[13] Consultations indicated higher total support for seeing demonstration projects as most important (53 per cent) as against basic and applied research (44 per cent). However, the consultations had a built-in bias, with a far higher number of respondents representing the industry community than the research community (42 per cent vs 20 per cent).

5.3   EU MARKET-PULL POLICIES

In December 2008, the EU adopted its climate and energy policy package for achieving the 20-20-20 targets by 2020, as a first step towards a low-carbon economy by 2050. Included were two cross-sector policy instruments: the revised EU ETS Directive and the Effort Sharing Decision (ESD) specifying binding emissions reduction targets for ETS and non-ETS sectors, respectively.[14] Also included were two 'technology-specific' instruments: the Renewable Energy Directive (RED), setting binding national targets for share of renewables, and the CCS Directive.[15] This package of EU climate and energy policies was implemented from 2009 onwards, with highly differing experiences.

5.3.1   *The CCS Directive*

This Directive (2009/31/EC) established a legal framework for safe geological storage of CO_2 and determined that new modernized coal plants should be made CCS-ready. It did not oblige the member-states to realize CCS but aimed at providing legal clarity to investors in the power sector. The ETS-generated carbon price was envisaged as main market-pull instrument for achieving the EU CCS targets: up to 12 pilot plants by 2015, with the aim of large-scale demonstration linked to coal by 2020.

Implementation of the CCS Directive encountered major challenges in several member-states. In Poland, implementation was seriously delayed (Skjærseth 2014, p. 18); in Germany and the Netherlands, public opposition restricted underground storage of CO_2 (Skjærseth et al. 2016). Such implementation problems led investors to doubt whether CCS could become a promising low-carbon solution—and most pilot projects already planned by industry were postponed or cancelled.

5.3.2   *EU ETS Directive*

This Directive aimed at 'pulling' the market for low-carbon energy technologies by means of a carbon price that would grow stronger with

---

[14] ETS sectors should reduce emissions by 21 per cent below 2005 levels. The ESD set different binding national targets, intended to yield a 10 per cent reduction for sectors not covered by the ETS.

[15] New policies on energy efficiency, fuel quality and car emissions were proposed according to a different schedule.

increasing scarcity of allowances in the market. It also stipulated that at least 50 per cent of public revenues from allowance auctions should target investments in low-carbon activities (Article 10.3), including the SET-Plan priorities. Additionally, funds were to be provided for the EU NER 300 Programme, in order to accelerate two SET-Plan priorities: CCS and renewable energy technology demonstration projects.

The ETS Directive was intended to drive a higher carbon price, in 2008 anticipated to reach €20–30/ton CO_2 by 2020. Reality, however, proved quite different. From 2008 onwards, the market experienced an increasing mismatch between supply and demand of allowances. Demand plunged, mainly due to downscaled industrial activity generated by the economic crisis and fuel-switching to renewable energy. The result was a plunge in the carbon price, from nearly €30 in spring 2008 to just above €5 from 2012 to 2018.

Besides providing poor market-pull signals, the low carbon price affected SET-Plan implementation more directly, by contracting the level of public revenues available for the funding of the priorities of the Plan. This included a downscaled NER 300 fund, which contracted from the expected €9 billion to €2.1 billion (Åhman et al. 2018). Most affected were the largest-scale demonstration projects, such as CCS projects, as NER 300 was to distribute funds evenly among the member-states and various technologies. Despite lower-than-expected budgets, the member-states still earned massive funds from auctioning, close to €12 billion between 2013 and 2015 (Ecologic 2016). Some 85 per cent of this income has been spent as intended, mainly on renewables, energy efficiency and cross-cutting climate policy programmes.

Overall, then, the EU ETS failed to provide carbon prices high enough to incentivize industry investments in SET-Plan demonstration projects for low-carbon technologies. This affected CCS demonstration projects, as they relied on the carbon price as well as ETS funding of the NER 300 Programme.

5.3.3   The Renewable Energy Directive

This Directive was specifically aimed at 'pulling' the market for renewable energy technologies, including those selected for the SET-Plan, through binding national targets that would be aggregated to 20 per cent renewable energy consumption in the EU collectively by 2020.[16] The RED would strengthen renewable energy policies already existing within the EU: most member-states had renewable energy deployment policies in place (feed-in tariffs and quota systems), with suitable demonstration projects in the pipeline (Åhman et al. 2018). Implementation of the Renewable Energy Directive has largely progressed in line with intentions. In 2014, the EU achieved a 16 per cent share and an estimated 16.4 per cent share in 2015, compared to 7 per cent in 2005. Most member-states are now on track to meet their 2020 targets (Skjærseth et al. 2016; Commission 2017b).

The combination of binding national renewable energy targets and stepped-up support schemes in the member-states became a key driver for realizing the part of the SET-Plan that focused on technologies for renewable electricity generation. Rapid and massive investment in wind and solar power technologies brought about the desired reductions in technology costs. However, this also had unintended effects on the electricity market. Wholesale prices started on a downward trend, with lower margins for fossil-power plants and growing fears that large-scale coal plants might become redundant, to be replaced by alternative decentralized renewable energy technologies (Auer and Haas 2016). These electricity market effects added risks to investing in CCS demonstration projects.

For the transport sector, the RED established a 10 per cent minimum binding target for renewable energy, replacing previous policies specifically aimed at stimulating market uptake of biofuels. This change reflected increasing doubts on the use of biofuels as regards sustainability and the effects on global food production (Di Lucia 2013). The new legal framework caused uncertainty for investors in biofuels due to several unresolved questions: which biofuels, and from which feedstock, should be accounted as 'sustainable'? How should the climate effects of biofuels be calculated, and what support schemes should be allowed to trigger investments in the 'right' biofuels? Despite the adoption of the Directive on Indirect Land-use Changes in 2015, the political and long-term market uncertainties persisted due to sluggish implementation of biofuel policies in the

---

[16] The targets were based on a flat rate of 5.5 per cent, with the remaining gap allocated according to GDP/capita. Targets among EU members range from 10 per cent in Malta to 49 per cent in Sweden.

member-states[17]—with negative effects on large-scale advanced biofuel projects (Åhman et al. 2018).

5.3.4   Summary

The context of climate and energy policies can help to explain the mixed experiences with SET-Plan implementation. The implementation of market-pull policies has varied roughly in line with the relative performance of the European Industrial Initiatives (EIIs). In part, the failure of industry investments in CCS demonstration projects was affected by EU policies that generated insufficient legal certainty for investors and incentives due to low carbon prices resulting from the EU ETS and by effects in the electricity market that gave rise to doubts about the future of coal-based power. The relatively good performance of certain renewable electricity technology demonstration projects (wind and solar technologies) followed from implementation of the Renewable Energy Directive, which created strong demand and certainty for investors. By contrast, the sustainability provisions of this Directive shifted, creating uncertainties for investments in demonstration of advanced biofuel technologies.

5.4   Conclusions

The challenges to strengthen, focus and give coherence to low-carbon energy research and innovation along the lines of the initial SET-Plan can largely be explained by a combination of analytical approaches. In line with Liberal Intergovernmentalism, diverse member-state R&I interests and priorities can partly explain the expansion in the number of prioritized SET-Plan technologies. Further, member-states lacked commitment to the SET-Plan, and many did not participate actively in the Plan's governing bodies.

Implementation challenges can be further explained from a Multilevel Governance perspective. The Commission played an active role in implementing the SET-Plan, but lacked authority to decide on funding in line with Plan priorities. Budget cycles were not synchronized with the SET-Plan, and within the Commission itself there were diverging interests that weakened its influence. The European Parliament had co-funding

---

[17] Most EU countries have come about halfway to their 2020 targets. In some member-states, the market share for biofuels has been declining (Skjærseth et al. 2016).

authority over R&I but had other priorities. Industry was not sufficiently committed to the EIIs because of scattered public funding and risk aversion aggravated by the financial crisis. The research community was internally split on technology priorities, but preferred longer-term basic research. Only to a limited degree did the research community commit to coordinate its activities with industry on the EII arena. Variation in EII performance has been closely related to variation in implementation of EU market-pull policies.

In retrospect, we can see that the SET-Plan was in many ways already doomed when it was made because of its governance system. A new structure for strengthening, focusing and giving coherence to low-carbon energy research was established—based on built-in tensions that were never resolved. And, responsibility for overseeing the Plan did not match authority to decide on funding: funding authority was given to decision makers that had other priorities.

REFERENCES

Advisory Group on Energy. (2008). *Advice on Issues Relating to the 2009 Work Programme, Based on Group Meeting 13 February 2008.* https://ec.europa.eu/research/fp7/pdf/old-advisory-groups/energy-wp-2009.pdf. Accessed 26 June 2018.

Åhman, M., Skjærseth, J. B., & Eikeland, P. O. (2018). Demonstrating Climate Mitigation Technologies: An Early Assessment of the NER 300 Programme. *Energy Policy, 117,* 100–107.

Auer, H., & Haas, R. (2016). On Integrating Large Shares of Variable Renewables into the Electricity System. *Energy, 115,* 1592–1601.

Commission. (2008, July). *European Community Steering Group on Strategic Energy Technologies: Terms of Reference.* Brussels.

Commission. (2009a). *SETIS Summary of Roadmaps.* https://setis.ec.europa.eu/summary-of-roadmaps. Accessed 13 June 2018.

Commission. (2009b). *Investing in the Development of Low Carbon Technologies (SET-Plan).* COM(2009) 519 Final, 07 October. Brussels: European Commission.

Commission. (2011). *Summary of the Outcomes of the Meeting of the Steering Group on Strategic Energy Technologies,* 27 October. Brussels, 14 November 2011.

Commission. (2012a). *Summary of the Outcomes of the Meeting of the Steering Group on Strategic Energy Technologies,* 02 July. Brussels, 03 July 2012.

Commission. (2012b). *Summary of the Outcomes of the Meeting of the Steering Group on Strategic Energy Technologies,* 17 September. Brussels, 20 September 2012.

Commission. (2013a). *Review of the SET-Plan Implementation Mechanisms for the Period 2010–2012*. Petten: Joint Research Centre of the European Commission.

Commission. (2013b). *Report of the Public Consultation on the Communication on Energy Technologies and Innovation, DG Energy*, 29 April. Brussels: European Commission.

Commission. (2015). *Towards an Integrated Strategic Energy Technology (SET) Plan: Accelerating the European Energy System Transformation*. COM (2015) 6317 Final. Brussels.

Commission. (2016). *Innovative Financial Instruments for First-of-a-Kind, Commercial-Scale Demonstration Projects in the Field of Energy*. Report Written by ICF in Association with London Economics, September 2016. Brussels: European Commission.

Commission. (2017a). *Interim Evaluation of Horizon 2020*. Commission Staff Working Document, SWD (2017) 221 Final, 29 May. Brussels: European Commission

Commission. (2017b). *Renewable Energy Progress Report*. COM (2017) 57 Final, 01 February. Brussels: European Commission.

Council. (2010). *Council Conclusions on the Commission Communication 'Investing in the Development of Low Carbon Technologies.'* Council of the European Union, 3001st Council Meeting, 12 March. Brussels: Council.

Council. (2013). *Conclusions on Smart Cities and Communities—European Innovation Partnership*, 3233rd Environment Council Meeting, 21 March. Brussels: Council.

Di Lucia, L. (2013). Too Difficult to Govern? An Assessment of the Governability of Transport Biofuels in the EU. *Energy Policy, 63*, 81–88.

Ecologic. (2016). *Smart Cash for the Climate: Maximising Auctioning Revenues from the EU Emissions Trading System*. An Analysis of Current Reporting by Member States and Options for Improvement. Brussels: Ecologic.

EERA. (2010, October). *FP8 Position Paper of the European Energy Research Alliance (EERA)*. http://ec.europa.eu/research/horizon2020/pdf/contributions/prior/eera.pdf. Accessed 13 June 2018.

EERA. (2018). *EERA Map of Members*. https://www.eera-set.eu/map/. Accessed 24 June 2018.

European Parliament. (2010a). Motion for a Resolution Further to Questions for Oral Answer B7-0011/2010 and B7-0012/2010 Pursuant to Rule 115(5) of the Rules of Procedure on Investing in the Development of Low Carbon Technologies (SET-Plan), Pilar del Castillo Vera, Christian Ehler, Herbert Reul, Maria da Graça Carvalho, Romana Jordan Cizeljo, on Behalf of the PPE Group; Giles Chichester on Behalf of the ECR Group, B7-0149/2010, 03 March.

European Parliament. (2010b). Motion for a Resolution Further to Questions for Oral Answer B7-0011/2010 and B7-0012/2010 Pursuant to Rule 115(5) of the Rules of Procedure on Investing in the Development of Low Carbon

Technologies (SET-Plan), Teresa Riera Madurell on Behalf of the S&D Group; Fiona Hall on Behalf of the ALDE Group; Claude Turmes on Behalf of the Verts/ALE Group, B7-0148/2010, 3 March.

European Parliament. (2010c). *Debates Thursday, 11 March 2010: Investing in Low-Carbon Technologies.* CRE 11/03/2010–2. Strasbourg: European Parliament.

European Parliament. (2010d). *European Parliament Resolution of 11 March 2010 on Investing in the Development of Low Carbon Technologies (SET-Plan).* Strasbourg: European Parliament.

European Parliament. (2012). *Report on the Proposal for a Regulation of the European Parliament and of the Council Establishing Horizon 2020—The Framework Programme for Research and Innovation (2014–2020).* Committee on Industry, Research and Energy. Rapporteur: Teresa Riera Madurell, A7–0427/2012, 20 December.

European Parliament. (2013). *European Parliament Legislative Resolution of 21 November 2013 on the Proposal for a Council Decision Establishing the Specific Programme Implementing Horizon 2020—The Framework Programme for Research and Innovation (2014–2020),* P7_TA-PROV (2013)0504.

JRC. (2015). *Capacity Mapping: R&D Investment in SET-Plan Technologies* (JRC Science and Policy Report). Petten: Joint Research Centre of the European Commission.

Research Council of Norway. (2011). *Kort om SET-Planen (SET-Plan Brief).* Internal Note, March 2011, revised 2013. https://www.forskningsradet.no/servlet/Satellite?blobcol=urldata&blobheader=application%2Fpdf&blobheadername1=Content-Disposition%3A&blobheadervalue1=+attachment%3B+filename%3DKortomSET-planen.pdf&blobkey=id&blobtable=MungoBlobs&blobwhere=1274502878282&ssbinary=true. Accessed 12 June 2018.

Skjærseth, J. B. (2014). *Implementing EU Climate and Energy Policies in Poland: From Europeanization to Polonization?* (FNI Report 8/2014). Lysaker: Fridtjof Nansen Institute.

Skjærseth, J. B. (2018). Implementing EU Climate and Energy Policies in Poland: Policy Feedback and Reform. *Environmental Politics,* 1–20. https://doi.org/10.1080/09644016.2018.1429046.

Skjærseth, J. B., Eikeland, P. O., Gulbrandsen, L. H., & Jevnaker, T. (2016). *Linking EU Climate and Energy Policies: Policymaking, Implementation and Reform.* Cheltenham: Edward Elgar.

CHAPTER 6

Conclusions and Prospects

The 2008 EU Strategic Energy Technology Plan (the SET-Plan) responded to the decline in funding for low-carbon energy research, scattered resource allocation and lack of acceleration and deployment of new technologies. Few resources were allocated to projects in the critical phase of demonstrating technologies in full scale. At the ten-year anniversary of the SET-Plan, this book has taken stock of its emergence and development: How was it established? Have implementation and performance been in line with intentions? How to explain the making and implementation of the SET-Plan?

Let us briefly recapitulate the basic features of the Plan. The aim was for the research community, industry, member-states and EU institutions to raise and pool their resources for the development of a limited range of technologies, to be promoted by large-scale demonstration projects beyond the capacity of any single country. The result was to be the acceleration of low-carbon energy innovation and deployment in the EU, in order to attain climate and energy policy targets and strengthen the EU's position in global energy-technology markets.

When the Commission proposed the SET-Plan in 2007, the focus was to be on solar, wind, nuclear fission, bioenergy/biofuels, CCS and electricity grid infrastructure. This selection was biased towards energy-supply technologies, primarily electricity supply. The Plan included a new governance system for defining, funding and coordinating joint projects for the

© The Author(s) 2020   117
P. O. Eikeland, J. B. Skjærseth, *The Politics of Low-Carbon
Innovation*, https://doi.org/10.1007/978-3-030-17913-7_6

new priorities, by designating specific roles to the research community, industry, member-states and the EU institutions in several new arenas.

Industries should take the lead in developing European Industrial Initiatives (EIIs), specifying the resources needed for full-scale demonstration projects for the six selected technologies and agreeing on funding together with national and EU public funding agents. As two-thirds of the total energy research and innovation resources in the EU originated from private industrial companies, they would be pivotal for developing large-scale demonstration projects that could prepare the way for actual market deployment. Public co-funding from national and EU programmes would increase the chance of realizing demonstration projects by reducing the risks of industry engagement. Research institutions were given the lead role in a European Energy Research Alliance (EERA) to develop joint research programmes aligned to the needs of the industrial initiatives.

The Commission and SET-Plan member-states were given the overarching roles for implementing the Plan. Under Commission leadership, national representatives would meet regularly in the SET-Plan Steering Group to discuss and decide implementation of the Technology Roadmaps for the EIIs and joint EERA programmes and pooling of funds towards these priorities. The Commission, for its part, was to ensure coherence between the SET-Plan and EU funding programmes. These programmes were managed by various Commission services, but were decided by the Council and the European Parliament. The Commission's Joint Research Centre would develop and administer a European energy technology information system (SETIS). This reporting, monitoring and review system would keep track of progress in implementing the Plan at EU, national and industry levels. Updated overviews of R&I funding and project development were needed to plan the allocation of resources to SET-Plan priorities.

The SET-Plan was adopted simultaneously with a larger EU climate and energy policy package for attaining the 2020 targets. This package included carbon pricing based on the EU ETS, binding national renewable energy targets, binding national targets for sectors not covered by the EU ETS and rules for safe geological storage of CO_2—intended to lay the ground for the construction of 12 pilot plants by 2015, with full-scale demonstration by 2020.[1] The wider policy package would enable implementation of the SET-Plan by providing for market-pull stimuli for the

[1] A new Energy Efficiency Directive was adopted in 2012.

low-carbon technology priorities, thus reducing risks and incentivizing industry to realize the large-scale demonstration projects. By 2008, the EU had put in place an unprecedented package of policies aimed at pushing and pulling the development and deployment of low-carbon technologies.

6.1   FINDINGS

Chapter 3 analysed the development of the SET-Plan, from its establishment to implementation. Before the Plan was initiated, public and private energy research funding had been declining. Energy research programmes at national and EU levels were extremely heterogeneous and poorly coordinated. The EU lacked a specific steering focus for its policies to 'push' energy-technological development, and EU policies to 'pull' demand for such technologies towards market deployment were weak. The Commission acknowledged these challenges from 2000, taking steps towards a more strategic joint EU approach that included energy research and innovation through the European Research Area and European Technology Platforms (ETPs).

In 2006, the Commission launched the idea of the SET-Plan, followed by specific proposals in 2007. These included the new SET-Plan arenas: The Steering Group; SETIS for reporting, monitoring and review; the European Industrial Initiatives, to focus on large-scale low-carbon demonstration projects; and the European Energy Research Alliance to align national research to the priorities of the SET-Plan. Initially, the Commission aimed at 'mixed' governance, whereby competence for picking the winners would be delegated to the member-states based on the Commission's first selection of six priorities. Criteria for technology selection included EU added value/additionality; the willingness of actors (member-states and/or industry) to join forces; the potential market penetration of the technology in various time-horizons; and potential contribution to CO_2 reduction, security of energy supply, and competitiveness. However, it was not clear which criteria had been decisive for picking the individual areas of technology, and prioritized technologies immediately became subject to mobilization and political conflict. Excluded priorities and stakeholders pushed for the inclusion of other criteria and technologies according to their own values and interests.

The Commission proposal reflected a preference for greater competence for the EU level, to be based on jointly agreed strategic priorities and extended coordination. In contrast, the Council essentially wanted to maintain competence for the SET-Plan with national opportunities to select priorities according to national energy policy and R&I interests. The Council's position implied that implementation would be governed by voluntary coordination and not on a legal basis, restraining the Commission to apply only soft measures. Both the Council and the European Parliament contested the Commission's initial selection of priority technologies for the Plan, wanting focus on a broader portfolio of technologies. This ran counter to the core idea of concentrating public and private resources for a limited set of technology areas and thereby increasing the chances of successful innovation in these areas.

The Commission's opportunity to fund the SET-Plan initiatives by programmes at the EU level was also restricted. The Council and Parliament insisted on retaining full decision-making power over future funding programmes. However, the Council conclusions and Parliament resolution on the SET-Plan were not transferred into a final authoritative compromise text as with 'standard' EU legislation. The SET-Plan gained backing from the Council and European Parliament in 2008 but contained fundamental unsettled and equivocal signals of future priorities and governance of implementation.

Implementation of the SET-Plan commenced from 2008. The Commission set up all the new arenas for collaboration where member-state governments, industry and the research community should coordinate together with the Commission to deliver on the objectives of focusing, strengthening and giving coherence to EU energy research and innovation efforts. Focusing the Plan along the initially selected six technology areas soon failed. In 2009, a first new 'Smart Cities' initiative was added, to benefit local administrations based on demonstrating energy-efficient end-use technologies—a technology area initially excluded from the Plan. Revision in 2015 brought in many new priorities, without discarding those already in the Plan, although CCS and nuclear were down-prioritized. The Plan expanded to include technologies and solutions for 14 energy sectors by 2017. Efforts were no longer to be concentrated on developing single energy-supply technologies: the focus should be on solutions needed to develop the entire energy system, including energy storage and energy-efficient solutions for end-use sectors like transport, buildings and industry. Additionally, energy-supply technologies for development and demonstration were expanded to include ocean and geothermal energy and technologies to promote renewable and efficient heating and cooling.

While total EU funding of low-carbon energy R&I increased after the SET-Plan was decided, strengthening the funding for several of the Plan's initial priorities proved difficult. This was especially the case with funding the European Industrial Initiatives and the many large-scale demonstration projects planned for implementing these. The Steering Group and the other arenas were largely unable to align and coordinate major EU funding programmes for low-carbon energy R&I, like the energy part of Horizon 2020, the European Energy Program for Recovery and the NER 300 programme, so as to create the 'critical mass' of resources needed to advance all the industrial initiatives according to implementation plans. Coherence problems between EU-level programmes and between EU-level and national energy research programmes have persisted.

Thus, resources for leveraging industry funding became scattered and insufficient to trigger the necessary industry investments in large-scale demonstration projects, although there was some variation among the technology areas covered by the EIIs. Wind and solar PV demonstration projects have progressed reasonably well towards final investment decisions by industry; and several demonstration projects have been initiated under the Smart Cities and Communities Initiatives. The EII on bioenergy redirected its focus from biofuels to biomass-based heat and power; industry final investment decisions progressed well for the latter, whereas several advanced biofuels demonstration projects have failed to achieve final investment decision. The Electricity Grid Initiative has noted some successful demonstration projects. The EII on sustainable nuclear power postponed construction of demonstration projects towards 2030. And the CCS Industrial Initiative, with its programme for pilots and large-scale demonstration, proved a complete failure.

In Chap. 4, we sought to explain the making of the SET-Plan, examining the role of the member-states, EU institutions and non-state actors with international technological markets as a contextual factor. The SET-Plan had not been requested by the member-states: it was initiated independently, by the Commission (see below). On the other hand, the proposed governance system largely reflected member-state preferences for retaining control over competence in selecting priorities for and steering public R&I resources in line with national energy goals. The technology areas that were initially selected largely (but not fully) mirrored priorities in the major energy R&I-intensive member-states. Notably, the

Commission had excluded energy efficiency, which ranked high on the R&I agendas of many member-states.

The Commission prepared the ground, launched the idea and initiated the SET-Plan as an integral part of the wider EU climate and energy package. Input to the selection of priority technologies came from industry and the research community organized in the existing technology networks, the European Technology Platforms. One selection criterion applied by the Commission was that relevant actors should have already mobilized networks and defined strategic research agendas for specific technologies. This was contested by interests that were less connected to existing R&I networks.

Some low-carbon technology areas, notably various energy-efficiency technologies, had backing from industry groups and the research community, but nevertheless initially failed to become selected by the Commission—an omission that was not accepted by the Council and European Parliament as decision makers. The Climate Alliance—composed of hundreds of municipalities across Europe—also gave weight to the inclusion of energy efficiency as priority for the SET-Plan and the Commission's specific selection of the 'Smart Cities' initiative. DG Research, the research community and the European Parliament pushed for including less-mature energy-supply technologies in a longer time frame. Many of these were added to the Plan as part of the reform in 2015.

An important driver for EU R&I efforts in general and the SET-Plan was the wish to strengthen the competitiveness of EU industries in international technology markets. Therefore, we had expected the initial selection of technologies to reflect areas for which European industry already held a dominant international niche position. The SET-Plan prioritized some technologies where EU companies had already established leadership but also technologies where other countries were leaders at the time, like solar PV and nuclear fission. The Plan's priorities were motivated mainly by the need to develop the internal energy market and to forge political agreement on the new EU climate and energy policy package, in order to achieve the goals adopted for 2020.

Chapter 5 focused on explaining the challenges to implementation. Why did the EU largely fail in strengthening, focusing and giving coherence to low-carbon energy research along the lines prescribed by the SET-Plan? We examined the role of the member-states, EU institutions, non-state actors and EU market deployment policies as contextual factors. Diverse member-state commitments to the Plan and diverging national interests and R&I priorities can help to explain the implementation challenges that arose. First, most member-states lacked commitment to the Plan, and many did not participate actively in the arenas established for governing the Plan. Second, many member-states did not prioritize low-carbon R&I in line with the SET-Plan—national programmes were focused more widely on other low-carbon technologies as well. Third, national competence over public research and innovation prevailed and contributed to the expansion of the SET-Plan. National competence, diverse interests and lack of commitment made it difficult for the Commission to achieve its intentions for the SET-Plan.

The potential for extended R&I cooperation on the SET-Plan priorities among member-states grew because of increased total funding, but this potential remained largely untapped. This untapped potential was not due solely to the lack of commitment by member-state governments: it also reflected governance challenges at the EU level. The Commission had a leading role in promoting implementation, by facilitating cooperation in the new arenas, but it lacked decision-making authority over EU-level R&I programmes so as to fill funding gaps and harmonize eligibility criteria. Budget cycles were not synchronized with SET-Plan decision making, and the Commission itself was internally split on the Plan's priorities. Internal diversity made the Commission vulnerable to pressures and demands from other stakeholders and weakened its efforts to align the SET-Plan with EU funding programmes.

Differing interests within the Commission reflected services with roles and tasks that catered to a range of stakeholder interests. DG Energy catered first and foremost industry interests that wanted priority for large-scale demonstration projects of relatively mature technologies. DG Research accommodated all sectors of society, as well as the entire research community, which saw basic and applied research into many less-mature technologies as essential to enable a long-term transition of the energy system. Industry commitment became weak because of scattered and insufficient public co-funding, risk aversion aggravated by the financial crisis, and low industry commitment. The diversified research community largely failed to link up to industry to advance joint projects.

The problems facing the Commission were exacerbated by the European Parliament, which had no formal role in the SET-Plan governance system but had decision-making authority over EU R&I programmes. The Parliament increasingly pressed for the future EU energy system to be based on technology priorities not initially focused by the SET-Plan, including small-scale decentralized technology, and sided with the research community/DG Research in support of the long-term development of less-mature technologies. Compromises after deliberations among MEPs resulted in support for the initial SET-Plan priorities and for the addition of new priorities and criteria.

Finally, an important contextual factor that affected implementation of the SET-Plan concerned climate and energy market-pull policies that came to create differing deployment conditions for different technologies, which in turn helps to explain variation among the EIIs in realizing large-scale demonstration projects. Such projects would be regarded as risky for private investors if market-pull policies and deployment conditions were weak or unstable. The CCS failure was caused partly by lack of legal certainty for investors and the failure of the EU ETS to generate higher carbon prices. The Renewable Energy Directive proved instrumental for creating long-term stability of demand, reducing commercial risks for industry. Variation among renewable technologies reflected technology-specific feed-in tariff systems biased in many countries towards ambitious market support for windpower, solar PV and biomass-based heat and power. Concerning biofuels, the policy signals at EU and national levels on sustainability were shifting and unstable (Åhman et al. 2018). In important member-states, future markets for nuclear power were uncertain, because of political decisions for phase-out.

The first analytical implication of these empirical findings relates to the complementary nature of the Liberal Intergovernmentalism (LI) and Multilevel Governance (MLG) approaches (see Chap. 2). These perspectives point to different mechanisms and explain different elements of making and implementing the SET-Plan. LI offers a good explanation of why SET-Plan governance became decentralized, based on national self-determination over the energy mix and related public R&I resources. It also explains how and why initial prioritized technologies tended to reflect R&I priorities in the major member-states. Diversity in R&I priorities and decentralization are important explanations for the implementation problems witnessed: expansion of the initial technology priorities, lack of coherence in funding, and inability to join forces for the realization of joint research and innovation programmes.

MLG complements LI by explaining how the Commission independently took the initiative to the SET-Plan and decided the initial priorities on the basis of existing industry and research community networks. The Commission developed the Plan as an integral, mutually dependent part of a wider package of EU climate and energy policies. This integration placed pressure on member-states to adopt all the policies, including the SET-Plan, simultaneously. MLG further helps to explain why the Commission was unable to resist pressures to expand the Plan. The Commission itself had a range of R&I interests vested with various services, making it susceptible to conflicting demands. Furthermore, the European Parliament—which had funding authority and technology priorities different from those of the Commission—put its imprint on EU funding programmes that affected implementation of the SET-Plan.

Although this study has brought us closer to understanding the development of the SET-Plan, there is still room for alternative explanatory perspectives and approaches to contribute to a deeper understanding of the challenges to implementation. First, domestic politics can help to explain preference formation in the field of energy research and innovation in SET-Plan member-states—for example, how various state–society relations and institutions linking state and society worked to establish R&I policies that promoted or delegitimized specific energy technologies like renewables, grids, CCS and nuclear power (Tsebelis 2002; Skjærseth et al. 2016). Whereas countries tend to prioritize R&I in line with their energy-mix and energy-dependence interests, changes may occur that affect or even redirect these relationships—for example, the Fukushima nuclear disaster that led Germany to reactivate its goal of phasing out nuclear power and to step up its efforts in deploying renewables and other energy sources (Eikeland 2016).

Second, we have seen that public and private actors share common interests across various levels of governance and EU organizational boundaries. For example, some groups have preferred funding of long-term basic and applied research at the expense of funding demonstration of technologies that are closer to market commercialization. Analysis of the strategies and roles of such groups in the making and implementation of the SET-Plan may profit from various approaches. One possibility would be the organizational fields approach—describing networks of non-state and related public actors bound together by frequent interaction based on what is deemed proper, natural or legitimate (DiMaggio and Powell 1991; Skjærseth and Eikeland 2013). Another possibility is the Advocacy Coalitions approach, for deeper analysis of how competing networks collaborate to promote specific low-carbon technology areas (Szarka 2010).

Finally, systems of innovation approaches could help to explain relationships and interactive learning between actors clustering within and between SET-Plan arenas. Systems of innovation depict the innovation process, from ideas to markets, as shaped by interactive learning among actors (companies, people) and institutions, and may be approached from different technologies or levels (international, national, regional or local system) (Edquist 1997). A focus on interactive learning among actors representing different innovation-chain functions might explain why some technology areas progressed differently from others and were either selected for the Plan or rejected (Bergek et al. 2015; Markad et al. 2015). A better understanding of the emergence and progress of the specific Smart Cities and Communities Initiative of the SET-Plan may be gained by applying conceptual frameworks developed in the literature on regional innovation systems (Cooke et al. 1997).

6.2   Lessons

Three key lessons can be drawn from this study. First, the SET-Plan was obstructed by a mismatch in competence between those with main responsibility for governing the Plan and those with authority to provide the resources necessary for implementation. The Commission was given the main role for governing and implementing the Plan, but lacked the formal authority to provide adequate resources. The member-states and the European Parliament had authority over the public resources needed, but they had other priorities and lacked commitment to the Plan. The result was low public budgets pooled for the technology priorities at national and EU levels, often insufficient to incentivize industry investments in large-scale demonstration projects.

A second, related, lesson is that the fundamental SET-Plan idea of strategically picking winners in the contest for scarce public R&I resources at EU level (in contrast to the technology neutrality principle) has proved futile without the transfer of corresponding competence to this level.

Finally, alignment with market-pull policies are crucial for realizing new low-carbon technologies. Broader climate and energy policies aimed to facilitate market uptake of the SET-Plan technologies and thus decrease the political risk for industry investments in demonstration projects. However, after their adoption in 2008 these policies progressed variously, resulting in widely differing market-adoption incentives for the various low-carbon technologies.

When reforming its climate, energy and technology policies towards 2030, has the EU been able to learn from these experiences? The revised

SET-Plan (Chap. 3) includes some minor changes in governance to spur greater commitment and more effective implementation, like new declarations signed by national governments and mergers of platforms for similar technologies. However, the revised Plan does not deal with the fundamental problem of competence mismatch in governing and funding the Plan.

The broader scope of the revised Plan—to fit better the diverse technology interests and funding realities among member-states and industries—will probably make the Plan less contested and might attract broader commitment to seeing it realized. However, there will also be new challenges as regards funding the Plan in its entirety—the amount of resources needed for realizing investments in all the new priorities has multiplied. Unless far greater R&I resources are committed, the resources may well remain scattered, not pooled in critical masses enough to accelerate large-scale demonstration projects. The initial strategic element of the SET-Plan may thus lose further ground.

6.3 Prospects

Since the SET-Plan reform in 2015, the Juncker Commission has taken additional steps aimed at improving implementation. First, the Commission in 2016 proposed a new mandatory regulation for full Energy Union governance of national integrated climate and energy plans that also included new governance for energy R&I and SET-Plan implementation (Commission 2016a). The Commission intended to strengthen EU governance of low-carbon energy R&I by a legal instrument binding for the member-states. Included was a provision that would oblige member-states to include in their national climate and energy progress reports information on objectives and policies for translating the SET-Plan to the national context; objectives for total (public and private) spending on R&I in clean-energy technologies and, as appropriate, long-term deployment targets and policies towards 2050.

However, the Council put forward amendments that restricted the transfer of governance competence from the member-states to the Commission. The time frame for meeting national R&I objectives, and how targets should reflect the priorities of the SET-Plan, were made conditional on such provisions being 'appropriate' (Council 2017). Provisions on reporting requirements were weakened: the member-states were merely to report private R&I efforts where such data were 'available'. In June 2018, the Council and Parliament reached an agreement on the

Governance Regulation, with the changes proposed by the Council intact (Council 2018). The Commission is likely to face continued problems in gaining access to the data necessary for more effective governance of SET-Plan implementation.

The Juncker Commission has also addressed the problem of poor industry-mobilization of energy investment funding. In 2014, Juncker proposed a new general investment plan to facilitate economic recovery—the European Fund for Strategic Investments (EFSI), combining Commission and EIB money to co-fund and trigger corporate investments. Initially, EFSI was adopted for the period 2015–2018, aimed at raising €315 billion; in 2017 it was extended to 2020, with a new and higher €500 billion target (European Parliament and the Council 2015). That there was no reference to the SET-Plan in the EFSI regulation underscores how the Plan had failed to evolve as the key focal point for EU energy research and innovation. However, the regulation emphasized low-carbon technology research and innovation projects as an important area for funding.

Preliminary assessments of the EFSI indicate that low-carbon and fossil-fuel-based projects alike (mainly natural gas) have been funded, with the chief focus on support to low-risk mature technologies (European Parliament 2018). However, from 2018, the Commission aims to direct the EFSI towards low-carbon projects in line with the Paris Agreement (see below) and towards riskier first-of-a-kind projects, in line with the SET-Plan.

Broader future opportunities for raising EU-level funding for low-carbon energy-technology R&I include the next Framework Programme (FP9) for 2020–2027 and a follow-up of the NER 300 programme under the EU ETS. For FP9—'Horizon Europe'—the Commission has proposed a budget of €100 billion (Commission 2018a). A European Innovation Council is to be established in addition to the existing European Research Council. Missions or 'moonlanding projects', indicating greater priority to high-risk innovation projects, are expected particularly within the new thematic 'global challenges' area under the umbrella of the UN Sustainable Development Goals. However, the final content and scale of FP 9 are yet to be decided by the Council and the European Parliament.

As part of the revised ETS for 2021–2030, the EU has adopted a follow-up demonstration programme to the NER 300—the Innovation Fund (Åhman et al. 2018). The new programme is to continue with emission allowances as the funding source, now expanded to 400 + 50 million allowances, with a potentially higher budget and thus new opportunities for selection of larger-scale projects. However, the size and stability of the actual budget will hinge on future carbon prices. The EU has agreed on a more ambitious cap, increasing the annual linear reduction factor (up from 1.74 per cent to 2.2 per cent), a Market Stability Reserve that will temporarily withdraw allowances from the market, and mechanisms for 'invalidating' or permanently cancelling surplus allowances. Whether the revised EU ETS will spur high, stable carbon prices remains uncertain, but the carbon price more than doubled shortly after the ETS reform. Other changes include a widening of the scope to energy-intensive industry projects, more support at an earlier stage in the project lifecycle, and possibly more projects for each member-state. The Innovation Fund is likely to promote more scale-limited projects, as will expanding the scope of the programme to include industry projects (Åhman et al. 2018).

What about the new market-pull policies for 2030? Will these policies attract member-states and industry to co-fund and realize low-carbon large-scale demonstration projects? The revised EU ETS has been discussed above, because of its dual role in providing technology push and demand pull. A higher carbon price will provide stronger incentives for companies to invest in emission-reducing low-carbon technologies in the sectors covered by the system: power production, energy-intensive industry and aviation. Moreover, member-states will increase their incomes from auctioning of revenues that can be used for supporting domestic R&I (see Chap. 5). The preliminary conclusion would thus be that gradual reduction of the allowance surplus can provide greater confidence in the ETS for creating stable investment signals for low-carbon demonstration deployment.

Beyond the revised ETS, the EU has not proposed any additional measures—or any new goals or visions, for that matter—to promote market uptake of CCS. All CCS pilots have been cancelled; this technology has been down-prioritized in the SET-Plan and there seem to be few chances of its being selected under the Innovation Fund. The focus for this fund has shifted from CCS to CCU—away from large-scale storage of captured carbon to utilization of carbon, including binding captured carbon in new products. This may provide a better business case for industry and pose fewer acceptance problems than those experienced in connection with underground storage of CO_2 in several countries (Skjærseth et al. 2016).

The new Effort Sharing Regulation (ESR) covers the non-ETS sectors for 2021–2030, including transport, agriculture, buildings and waste. Binding national targets, to be based on GDP/capita, will vary between 2005 stabilization and 40 per cent emissions cut by 2030. There will be no more import of external credits, but EU internal flexibility is enhanced significantly: banking and borrowing of annual emission allocations, transfer between member-states and between the ESR and the EU ETS, and new accounting rules for land use, and land-use change and forestry. The CEECs have also secured a 'safety reserve' of surplus emission from these sectors. The ESR provides a regulatory framework, but it is up to each member-state to adopt policies and measures for its non-ETS sectors. Although the ESR is likely to prove more constraining than existing policies for 2020, flexibility will reduce the regulatory pressure on member-states and the demand for low-carbon technology deployment in non-ETS sectors. The ESR has the potential to increase deployment of technologies linked to all non-ETS sectors with transport as a key focus. Transport represents nearly a quarter of EU GHG emissions,

As for new EU renewable energy policies, new state-aid guidelines in 2014 restricted national levels of market support. The EU agreed to remove guaranteed dispatch and feed-in-tariffs because national support systems had led to barriers to competition and trade across borders which counteracted the EU internal energy market. National support policies were adapted to the new state-aid guidelines; many member-states used this opportunity to cut down on support, which has led to higher political risk for investors, as already observed in decreasing investments in demonstration projects for some SET-Plan technologies (Commission 2016b).

Further, the European Council has decided to abolish binding national renewable energy targets. After first agreeing in 2014 on a modest 27 per cent target for the share of renewables by 2030, only slightly above the reference 'business as usual' scenario of 24 per cent, negotiations between the Council and Parliament in 2018 reached a compromise on the higher 32 per cent level, but with the target retained as non-binding for the member-states (Commission 2018b). Also agreed on are new sustainability standards for biofuels, aimed at further restricting the market-pull of fuels based on food-based feedstock and strengthening incentives for development of advanced non-food-based biofuels. Altogether, less ambitious market-pull instruments for renewables seem set to materialize, except possibly for biofuels.

The upshot is that there has been a strengthening of some market-pull policies, like the ESR and the EU ETS, but a weakening of others, like the RED and CCS. Whether this will add-up as stronger or weaker incentives for investing in low-carbon energy-technology innovation depends on how the new policies are implemented.

Internationally, quite dramatic changes have occurred in public R&I, low-carbon technology investments and installed capacity (JRC 2015; REN21 2016). The EU has been surpassed by China as to wind and solar power; and the international wind turbine market that seemed so promising for European original equipment manufacturers back in 2008 has been facing competition from new Indian and Chinese entrants (see Chap. 4). In 2015, the expanding turbine home market in China was supplied almost exclusively by Chinese—not European—manufacturers. For the first time, a Chinese company (Goldwind) led the ranking of turbine manufacturers in terms of installed capacity (JRC 2017). There has been significant growth also in India, the United States and Brazil. The EU has clearly not been able to consolidate its first-mover advantages in these areas through the SET-Plan. The case of energy storage shows how the EU is becoming a 'follower' in some technologies.[2] From 2014, the EU gave energy storage higher priority with the SET-Plan revision.

New low-carbon technology initiatives by Horizon Europe's global challenges and the European Fund for Strategic Investment's reference to the Paris Agreement indicate that the Commission is increasingly using the international context to increase energy R&I funding. In 2015, Mission Innovation was launched by 22 countries and the EU at the Paris Climate conference.[3] The aims mirror those of the original SET-Plan: to accelerate clean-energy innovation to bring down costs, double public clean-energy R&D investments by 2021, encourage private-sector investment, and cover the spectrum from early stage to technology demonstration projects. Each country will decide its own clean-energy R&D priorities, but seven actions (or 'innovation challenges') have been identified: smart grid, off-grid access for households and communities, carbon

---

[2] In 2011, China was the clear leader in energy storage R&D, followed by Japan, the EU and the United States. Solar and wind produce variable energy. Energy storage for geographically decoupling energy supply and demand has become a key emerging low-carbon technology. The focus is on reducing the costs of high-density storage and related development of thermo-chemical processes (JRC 2015).

[3] http://mission-innovation.net/. Accessed 13 June 2018.

capture, sustainable biofuels, converting sunlight into storable fuel, clean-energy materials, and affordable heating and cooling of buildings. Many of these priorities are now generally covered by EU energy research and innovation.

How the future for EU low-carbon energy research and innovation will look is highly uncertain. To ensure longer-term success in line with the 2050 decarbonization target and international energy research and innovation leadership, the EU will need to find a better way to combine technology developments from the market 'below' with competence to strategically prioritize at EU level and by the member-states. Based on input from industry and the research community, low-carbon technologies should match ambitious climate and energy market-pull policies.

The era of EU picking winners belongs apparently to the past. More technology-neutral policies will most likely reduce political conflict in Europeanization, but weak prioritization may also reduce prospects for decarbonization by 2050.

REFERENCES

Åhman, M., Skjærseth, J. B., & Eikeland, P. O. (2018). Demonstrating Climate Mitigation Technologies: An Early Assessment of the NER 300 Programme. *Energy Policy, 117*, 100–107.

Bergek, A., Hekkert, M., Jacobsson, S., Markard, J., Sanden, B., & Truffer, B. (2015). Technological Innovation Systems in Contexts: Conceptualizing Contextual Structures and Interaction Dynamics. *Environmental Innovation and Societal Transitions, 16*, 51–64.

Commission. (2016a). *Proposal for a Regulation on the Governance of the Energy Union.* COM (2016) Final/2, 23 February 2017. Brussels: European Commission.

Commission. (2016b). *Innovative Financial Instruments for First-of-a-Kind, Commercial-Scale Demonstration Projects in the Field of Energy.* Report Written by ICF in Association with London Economics, September 2016. Brussels: European Commission.

Commission. (2018a). *EU Budget: Commission Proposes Most Ambitious Research and Innovation Programme Yet.* Press Release, IP/18/4041. Brussels, 7 June.

Commission. (2018b). *Europe Leads the Global Clean Energy Transition: Commission Welcomes Ambitious Agreement on Further Renewable Energy Development in the EU.* Statement/18/4155. Strasbourg, 14 June.

Cooke, P., Uranga, M. G., & Etxebarria, G. (1997). Regional Innovation Systems: Institutional and Organisational Dimensions. *Research Policy, 26*(4/5), 475–491.

Council. (2017). *General Approach*, 12 December 2017 (OR. en) 15235/17, Interinstitutional File: 2016/0375 (COD). http://data.consilium.europa.eu/doc/document/ST-15235-2017-INIT/en/pdf. Accessed 26 June 2018.

Council. (2018, June 28). *Proposal for a Regulation on the Governance of the Energy Union – Analysis of the Final Compromise Text with a View to Agreement.* Brussels. http://data.consilium.europa.eu/doc/document/ST-10307-2018-ADD-2/en/pdf. Accessed 22 Oct 2018.

DiMaggio, P. J., & Powell, W. W. (1991). Introduction. In P. J. DiMaggio & W. W. Powell (Eds.), *The New Institutionalism in Organizational Analysis* (pp. 1–40). Chicago: University of Chicago Press.

Edquist, C. (Ed.). (1997). *Systems of Innovation: Technologies, Institutions and Organizations.* London: Pinter/Cassell Academic.

Eikeland, P. O. (2016). Implementation in Germany. In J. B. Skjærseth, P. O. Eikeland, L. H. Gulbrandsen, & T. Jevnaker (Eds.), *Linking EU Climate and Energy Policies: Policymaking, Implementation and Reform* (pp. 91–119). Cheltenham: Edward Elgar.

European Parliament. (2018, February 15). *European Fund for Strategic Investments – EFSI 2.0.* Briefing EU Legislation in Progress, Third Edition. http://www.europarl.europa.eu/RegData/etudes/BRIE/2016/593531/EPRS_BRI(2016)593531_EN.pdf. Accessed 26 June 2018.

European Parliament and the Council. (2015). Regulation (EU) 2015/1017 of 25 June 2015 on the European Fund for Strategic Investments, the European Investment Advisory Hub and the European Investment Project Portal. *OJ*, L 169/1, 01 July 2015.

JRC. (2015). *Capacity Mapping: R&D Investment in SET-Plan Technologies* (JRC Science for Policy Report). Petten: Joint Research Centre of the European Commission

JRC. (2017). *Wind Energy Status Report* (JRC Science for Policy Report). Petten: Joint Research Centre of the European Commission

Markad, J., Hekkert, M., & Jacobsson, S. (2015). The Technological Innovation Systems Framework: Response to Six Criticisms. *Environmental Innovation and Societal Transitions, 16,* 76–86.

REN21. (2016). *Renewables 2016 Global Status Report.* Renewable Energy Policy network for the 21st Century. Paris: REN21 Secretariat.

Skjærseth, J. B., & Eikeland, P. O. (Eds.). (2013). *Corporate Responses to EU Emissions Trading: Resistance, Innovation or Responsibility?* Farnham: Ashgate.

Skjærseth, J. B., Eikeland, P. O., Gulbrandsen, L. H., & Jevnaker, T. (2016). *Linking EU Climate and Energy Policies: Policymaking, Implementation and Reform.* Cheltenham: Edward Elgar.

Szarka, J. (2010). Bringing Interests Back in: Using Coalition Theories to Explain European Wind Power Policies. *Journal of European Public Policy, 17,* 836–853.

Tsebelis, G. (2002). *Veto Players: How Political Institutions Work.* Princeton: Princeton University Press.

APPENDIX: LIST OF INTERVIEWS

Interview A: Anger, Niels, DG Energy, Policy officer Unit A1; Brussels, 26.01.16.

Interview B: Bemtgen, Jean-Marie, DG Ener, Policy Officer for New Energy Technologies and Innovation Unit C2; Brussels, 26.01.16.

Interview C: Boneva, Melina, DG Climate Action, Policy Officer ETS, Policy and Auctioning Unit; Brussels, 25.01.16.

Interview D: Group interview with staff in DG Research and Innovation, Energy Directorate—Strategy Unit G1: Gwennael Joliff-Botrel, Head of Unit; Benedicte Caremier, Policy Officer; Thomas Schubert, Policy Officer; Brussels, 27.01.2016.

Interview E: Busuoli, Massimo, Head of NTNU Brussels office, member of the Secretariat of the European Energy Research Alliance (EERA) 2008–2015, and coordinator of EERA 2013–2015; Brussels, 29.01.16.

Interview F: Christensen, William, Norwegian Ministry of Petroleum and Energy, Director, Department of Climate, Industry and Technology; member of the SET-Plan Steering Group; Oslo, November 2014.

Interview G: Constantinescu, Norela, ENTSO-E, Senior Advisor Research & Innovation, former Policy Officer DG Energy, Energy Technologies and Research Coordination Unit; Brussels, 29.01.16.

Interview H: Egenhofer, Christian, Director and Associate Senior Research Fellow, CEPS; Brussels, 24.04.12.

Interview I: Eikaas, Tor Ivar, Research Council of Norway, Special Advisor and member of SET-Plan Sherpa Group; Oslo, September 2014.

© The Author(s) 2020   135
P. O. Eikeland, J. B. Skjærseth, *The Politics of Low-Carbon Innovation*, https://doi.org/10.1007/978-3-030-17913-7

Interview J: Helseth, Jonas, Bellona Europa, Director Brussels office, various chairing roles in Zero Emissions Platform and European Biofuels Technology Platform; Brussels, 29.01.16.

Interview K: Hodne, Tor Eigil, Senior Vice President European Affairs, Statnett SF; Brussels, 28.01.16.

Interview L: Kåberger, Tomas, Professor Industrial Energy Policy, Chalmers University, and Chair of European Biofuels Technology Platform Steering Committee; Oslo, 14.03.17.

Interview M: Medved, Karina, Eurelectric, Advisor Energy Policy and Generation Unit; Brussels, 27.01.16.

Interview N: Peteves, Efstathios, Head of 'Knowledge for the Energy Union' of the Energy, Transport and Climate Directorate of the European Commission's Joint Research Centre (JRC) and Leader of the Commission's Strategic Energy Technologies Information System (SETIS); Petten, the Netherlands, 27.02.16.

Interview O: Scott, Jesse, Eurelectric, Head of Unit for Environment and Sustainable Development; Brussels, 26.04.12.

Interview P: Slingenberg, Yvon, DG Climate Action, Director; Brussels, 28.01.16.

Interview Q: Støa, Petter, SINTEF Energy, Leader Brussels office; Brussels, 26.01.16.

Interview R: Tzimas, Evangelos, Deputy Head of the 'Knowledge for the Energy Union' Unit of the Energy, Transport and Climate Directorate of the European Commission's Joint Research Centre (JRC); Petten, the Netherlands, 27.02.16.

Interview S: Tulej, Piotr, Head of Unit Low Carbon Technologies, DG Climate Action; Brussels, January 2016.

Interview T: Kai Tullius, DG Energy, Policy Coordinator CCS Policy; Brussels, 23.04.12.

Interview U: Participation and discussions at NERA/NPUE meeting in Research Council of Norway with national representative of SET-Plan Steering Group, the Sherpa Group, Advisory Group to the Commission and various national SET-Plan stakeholders; Oslo, 16.12.15.

Index[1]

A

Advocacy Coalition approach, 125
Agenda-setting, 69
ALTENER programme, 19
Arenas in SET-Plan governance
   European Energy Research Alliance
     (EERA), 31, 118, 119; EERA
     Executive Committee, 50,
     106n12; EERA Secretariat, 50
   European energy technology
     information system (SETIS),
     29, 32, 34, 41, 49–52, 72,
     94–96, 98, 100, 101, 118, 119
   European Industrial Initiatives
     (EIIs), EII Teams, 30, 31, 34,
     50, 72, 96, 97, 99, 106,
     106n11, 107, 112, 118, 119,
     121
   Steering Group, 29, 31, 32, 34, 47,
     48, 66, 93–99, 93n1, 119, 121

B

Barroso Commission, 22n10
Berlin Model, 95
Bolzano Declaration, 80
Bottom-up, 7, 31, 65, 68

C

China, 29n22, 85–87, 131, 131n2
Climate Alliance, 80, 122
Climate change, 1, 18, 20, 22, 23, 26,
   38, 70, 72, 75, 80, 84, 93n1
Co-funding, 39, 40, 46, 52, 98, 99,
   101, 108, 112, 123
Competitiveness, 1, 12, 18–20, 22,
   23, 25n16, 31, 38, 40, 48n45,
   69, 70, 71n6, 72, 79, 82, 85,
   93n1, 119, 122
Covenant of Mayors, 80
Critical mass of resources, 46, 121

[1] Note: Page numbers followed by 'n' refer to notes.

© The Author(s) 2020   137
P. O. Eikeland, J. B. Skjærseth, *The Politics of Low-Carbon Innovation*, https://doi.org/10.1007/978-3-030-17913-7